W9-DBP-119

VHDL
Made Easy!

David Pellerin
Douglas Taylor

Prentice Hall PTR
Upper Saddle River, New Jersey 07458
http://www.prenhall.com

Library of Congress Cataloging-in-Publication Data

```
Pellerin, David.
     VHDL made easy! / David Pellerin, Douglas Taylor.
          p.     cm.
     Includes index.
     ISBN 0-13-650763-8
     1. VHDL (Computer hardware description language)  I. Taylor,
  Douglas.  II. Title.
  TK7885.7.P46  1996
  621.39'2--cd20                                    96-26999
                                                    CIP
```

Editorial/Production Supervision: Craig Little
Acquisitions Editor: Karen Gettman
Manufacturing Manager: Alexis R. Heydt
Cover Design: PM Workshop
Cover Design Director: Jerry Votta

© 1997 Prentice Hall PTR

Prentice-Hall, Inc.

A Simon & Schuster Company

Upper Saddle River, NJ 07458

All rights reserved. No part of this book may be
reproduced, in any form or by any means, without
permission in writing from the publisher.

All product names mentioned herein are trademarks of their respective owners.

The publisher offers discounts on this book when ordered in bulk quantities.
For more information, contact:

Corporate Sales Department
PTR Prentice Hall
One Lake Street
Upper Saddle River, NJ 07458
Phone: 800-382-3419, FAX: 201-236-7141
E-mail: corpsales@prenhall.com

Printed in the United States of America
10 9 8 7 6 5 4 3 2 1

ISBN 0-13-650763-8

Prentice-Hall International (UK) Limited, *London*
Prentice-Hall of Australia Pty. Limited, *Sydney*
Prentice-Hall of Canada, Inc., *Toronto*
Prentice-Hall Hispanoamericana S.A., *Mexico*
Prentice-Hall of India Private Limited, *New Delhi*
Prentice-Hall of Japan, Inc., *Tokyo*
Simon & Schuster Asia Pte. Ltd., *Singapore*
Editora Prentice-Hall do Brasil, Ltda., *Rio de Janeiro*

Contents

5. Understanding Concurrent Statements 141

6. Understanding Sequential Statements 163

Preface

Take a few dozen electronic design engineers at random and put them in a room. Now pose a question: "How many of you are using hardware description languages for your current design projects?" A small number of hands will go up.

Now ask another question: "How many of you think you will be using hardware description languages within two years?" The resulting blizzard of hands might surprise you.

HDL expertise is a critical, distinguishing skill, a skill you must have to succeed in your career as an electronics designer.

Why this sudden interest in HDLs? There are a number of factors, including a rapid increase in circuit complexities, an industry-wide desire for more formal (correct-by-design) engineering methods, and a general maturing of lower-cost, more accessible HDL tools.

Yet the biggest factor for the average engineer may be simple fear. Let's face it: the boom years of electronics are over — at least for the time being — and the era of cowboy engineering has passed. If you are an engineer who watches trends in the industry, you know that HDL expertise is a critical, distinguishing skill, a skill you must have to succeed in your career as an electronics designer.

So where do you get started? If you are not making use of HDLs in your current projects, how do you become familiar with HDL design concepts? How do you learn to use HDL tools when you have little or no budget to buy them?

Easy-to-follow descriptions of complex concepts, useful code samples, and a wealth of information...

The answer is here in this book and on the companion CD-ROM. Our goal in producing this book is to show you how HDLs — in this case VHDL (VHSIC Hardware Description Language) — can be used to describe and debug complex electronic circuits. (See the *Introduction* for more information on this acronym within an acronym.) In this book you'll find easy-to-follow descriptions of complex HDL concepts, useful VHDL code samples, and a wealth of information to help you get started with your own projects. You'll also find helpful tips and advice that will save you time as you progress on your way to becoming an expert HDL user.

What This Book Is

A practical tutorial on VHDL...

We'll be clear at the outset: this is not a reference manual on VHDL. There are enough books and specifications already in existence that describe the subtle nuances of VHDL syntax. Instead, this book is a practical tutorial on VHDL — one that focuses on concepts (such as hierarchy, sequential and concurrent descriptions, and test benches) that are common to other HDLs, including VHDL's main rival, Verilog HDL.

A second and equally important goal of this book is to introduce VHDL in the context of its most common use today: synthesis. In pursuit of this goal, we have minimized (in some cases eliminated) lengthy discussions about timing annotation and other issues of interest primarily to simulation model developers. Our assumption (verified through direct contact with hundreds of mainstream engineers) is that your first application of VHDL will be for synthesis, and you will therefore need to know how to write (1) design descriptions that are synthesizable, and (2) test benches to verify the correctness of those design descriptions.

Who Can Use This Book

This book is intended for experienced electronic design engineers and students of electronic design. Whether you are engaged in simple projects involving programmable logic devices or are developing large-scale ASICs (application specific integrated circuits), you will find the information in this book to be of high value.

Information that is accessible and enjoyable to read...

Although we have assumed a certain level of knowledge in the area of digital design and engineering fundamentals (for example, we assume you know how a flip-flop works), we have taken great pains to ensure that the information in this book is accessible and enjoyable to read. Rather than slow your progress with page after page of syntax diagrams and incomprehensible semantic rules, we present sample design descriptions which are intentionally brief, each designed to demonstrate a limited number of important HDL concepts.

Coding Styles

There seem to be as many styles of coding in VHDL as there are engineers using it. To make the examples in this book as readable as possible, we have used lower or mixed-case text when entering VHDL descriptions, and bold face type for all keywords in the source code figures.

We have not used any published style guide for such things as object names, ordering of statements, and line spacing. Instead, we have used whatever style seemed most appropriate for the example being presented. In your own projects, it makes sense to develop a consistent VHDL style and use that style throughout your company or project.

The Companion CD-ROM

The CD-ROM that accompanies this book includes demonstration software, example VHDL source files, and information useful to new and experienced VHDL users. The software on the CD-ROM is intended for use on Microsoft® Windows™–based personal computers and includes (among other things):

- Accolade Design Automation's *PeakVHDL*™ Simulator Demonstration Version 1.14.

- Capilano Computing System's DesignWorks™ Version 3.1.4 demonstration software.

- SynaptiCAD's The Waveformer™ Version 2.5 demonstration software.

- Visual Software Solution's StateCAD™ Version 2.10G demonstration software.

Acknowledgments

As anyone who has ever attempted to write a book knows, moving ideas from one's head to a bound volume is rarely a solitary process. This book is no exception. Many have contributed, and we are grateful to them all. Special thanks to Steven Lee of Hewlett Packard and Henry Lynn of Lockheed for their thoughtful and thorough technical reviews and comments. Thanks also to Edward Aung for providing the adder-subtractor example mentioned in Chapter 9, and to our Prentice Hall editor, Karen Gettman, for her tireless efforts—even in the midst of jury duty—to get this book to press.

Finally, and most of all, we thank Satomi and Kal, who patiently endured their husbands' all-too-frequent binges of writing and editing, glazed-out expressions at dinner, and unwillingness to get real jobs that made this book possible.

1. Introduction

VHDL makes it easy to describe very large circuits and systems.

VHDL (VHSIC Hardware Description Language) is becoming increasingly popular as a way to express complex digital design concepts for both simulation and synthesis. The power of VHDL makes it easy to describe very large digital designs and systems, and the wide variety of design tools now available make the translation of these descriptions into actual working hardware much faster and less error-prone than in the past.

The power and depth of VHDL does have its price, however. To make effective use of VHDL as a design entry tool, you must invest time to learn the language and, perhaps more importantly, to learn to use the higher-level design methods made possible by this powerful language.

Goals of This Book

It's easy to get started with VHDL ...

This book will introduce you to VHDL and show you how VHDL can be used to describe digital logic for implementation in programmable logic or ASICs. To this end, we avoid prolonged discussions appropriate only for developers of simulation models and system-level simulations. Instead, we concentrate on those aspects of the language that are most useful for synthesis. This book will give you enough information to quickly get started using VHDL and will suggest coding styles that are appropriate, using a wide variety of available synthesis and simulation tools.

Along the way, we will examine the many advantages of using VHDL for synthesis and simulation, and we will also explore some of the drawbacks of using VHDL when compared to alternative design capture methods. We will see how VHDL fits into the overall electronic design process, particularly as that process relates to FPGA, PLD and ASIC design problems.

In short, we will cover both VHDL syntax *and* how it is used for simulation and synthesis.

What Is VHDL?

VHDL is a programming language that has been designed and optimized for describing the behavior of digital systems. As such, VHDL combines features of the following:

A simulation modeling language

Use VHDL to create re-usable circuit building blocks...

VHDL has many features appropriate for describing (to an excruciating level of detail) the behavior of electronic components ranging from simple logic gates to complete microprocessors and custom chips. Features of VHDL allow electrical aspects of circuit behavior (such as rise and fall times of signals, delays through gates, and functional operation) to be precisely described. The resulting VHDL simulation models can then be used as building blocks in larger circuits (using schematics, block diagrams or system-level VHDL descriptions) for the purpose of simulation.

A design entry language

VHDL includes a rich set of control and data representation features.

Just as high-level programming languages allow complex design concepts to be expressed as computer programs, VHDL allows the behavior of complex electronic circuits to be captured into a design system for automatic circuit synthesis or for system simulation.

Like real hardware, VHDL allows concurrent events.

Like Pascal, C and C++, VHDL includes features useful for structured design techniques, and offers a rich set of control and data representation features. Unlike these other programming languages, VHDL provides features allowing concurrent events to be described. This is important because the hardware described using VHDL is inherently concurrent in its operation. Users of PLD programming languages, such as PALASM, ABEL, CUPL and others, will find the concurrent features of VHDL quite familiar. Those who have only programmed using software programming languages will have some new concepts to grasp.

A verification language

One of the most important applications of VHDL is to capture the performance specification for a circuit, in the form of what is commonly referred to as a test bench. Test benches are VHDL descriptions of circuit stimuli and corresponding expected outputs that verify the behavior of a circuit over time. Test benches should be an integral part of any VHDL project and should be created in tandem with other descriptions of the circuit.

A netlist language

While VHDL is a powerful language with which to enter new designs at a high level, it is also useful as a low-level form of communication between different tools in a computer-based design environment. VHDL's structural language features allow it to be effectively used as a netlist language, replacing (or augmenting) other netlist languages such as EDIF.

A standard language

Reuse design elements over and over.

One of the most compelling reasons for you to become experienced with and knowledgeable in VHDL is its adoption as a standard in the electronic design community. Using a standard language such as VHDL virtually guarantees that you will not have to throw away and recapture design concepts simply because the design entry method you have chosen is not supported in a newer generation of design tools. Using a standard language also means that you are more likely to be able to take advantage of the most up-to-date design tools and that you will have access to a knowledge base of thousands of other engineers, many of whom are solving problems similar to your own.

As simple or complex as you want

While it is true that VHDL is a large and complex language, it is not difficult to begin using. Start simple and explore advanced features as you become more confident. It won't be long before you are coding with the masters!

A Brief History of VHDL

First published as a standard in 1987, VHDL is a living standard that is reviewed every five years by a standards committee.

VHDL, which stands for VHSIC (Very High Speed Integrated Circuit) Hardware Description Language, was developed in the early 1980s as a spin-off of a high-speed integrated circuit research project funded by the U.S. Department of Defense. During the VHSIC program, researchers were confronted with the daunting task of describing circuits of enormous scale (for their time) and of managing very large circuit design problems that involved multiple teams of engineers. With only gate-level design tools available, it soon became clear that better, more structured design methods and tools would be needed.

To meet this challenge, a team of engineers from three companies — IBM, Texas Instruments and Intermetrics — were contracted by the Department of Defense to complete the specification and implementation of a new, language-based design description method. The first publicly available version of VHDL, version 7.2, was released in 1985. In 1986, the Institute of Electrical and Electronics Engineers, Inc. (IEEE) was presented with a proposal to standardize the language, which it did in 1987 after substantial enhancements and modifications were made by a team of commercial, government and academic representatives. The resulting standard, IEEE 1076-1987, is the basis for virtually every simulation and synthesis product sold today. An enhanced and updated version of the language, IEEE 1076-1993, was released in 1994, and VHDL tool vendors have been responding by adding these new language features to their products.

IEEE Standard 1164

IEEE 1076-1987 and 1164 together form the complete VHDL standard in widest use today.

Although IEEE Standard 1076 defines the complete VHDL language, there are aspects of the language that make it difficult to write completely portable design descriptions (descriptions that can be simulated identically using different vendors' tools). The problem stems from the fact that VHDL supports many abstract data types, but it does not address the simple problem of characterizing different signal strengths or commonly used simulation conditions such as unknowns and high-impedance.

Soon after IEEE 1076-1987 was adopted, simulator companies began enhancing VHDL with new, non-standard types to allow their customers to accurately simulate complex electronic circuits. This caused problems because design descriptions entered into one simulator were often incompatible with other simulation environments. VHDL was quickly becoming a nonstandard.

To get around the problem of nonstandard data types, another standard was developed by an IEEE committee. This standard, numbered 1164, defines a standard package (a VHDL feature that allows commonly used declarations to be collected into an external library) containing definitions for a standard nine-valued data type. This standard data type is called **std_logic**, and the IEEE 1164 package is often referred to as the standard logic package, or MVL9 (for multi-valued logic, nine values).

The IEEE 1076-1987 and IEEE 1164 standards together form the complete VHDL standard in widest use today. (IEEE 1076-1993 is slowly working its way into the VHDL mainstream, but it does not add significant new features for synthesis users.)

IEEE Standard 1076.3 (Numeric Standard)

IEEE Standard 1076.3 increases the power of Standards 1076 and 1164.

Standard 1076.3 (often called the Numeric Standard or Synthesis Standard) defines standard packages and interpretations for VHDL data types as they relate to actual hardware. This standard, which was released at the end of 1995, is intended to replace the many custom (nonstandard) packages that vendors of synthesis tools have created and distributed with their products.

IEEE Standard 1076.3 does for synthesis users what IEEE 1164 did for simulation users: increase the power of Standard 1076, while at the same time ensuring compatibility between different vendors' tools. The 1076.3 standard includes, among other things:

1) A documented hardware interpretation of values belonging to the **bit** and **boolean** types defined by IEEE Standard 1076, as well as interpretations of the **std_ulogic** type defined by IEEE Standard 1164.

2) A function that provides "don't care" or "wild card" testing of values based on the **std_ulogic** type. This is of particular use for synthesis, since it is often helpful to express logic in terms of "don't care" values.

3) Definitions for standard signed and unsigned arithmetic data types, along with arithmetic, shift, and type conversion operations for those types.

IEEE Standard 1076.4 (VITAL)

IEEE 1076.4 adds timing annotation to VHDL as a standard package.

The annotation of timing information to a simulation model is an important aspect of accurate digital simulation. The VHDL 1076 standard describes a variety of language features that can be used for timing annotation. However, it does not describe a standard method for expressing timing data outside of the timing model itself.

The ability to separate the behavioral description of a simulation model from the timing specifications is important for many reasons. One of the major strengths of Verilog HDL (VHDL's closest rival) is the fact that Verilog HDL includes a feature specifically intended for timing annotation. This feature, the *Standard Delay Format*, or *SDF*, allows timing data to be expressed in a tabular form and included into the Verilog timing model at the time of simulation.

The IEEE 1076.4 standard, published by the IEEE in late 1995, adds this capability to VHDL as a standard package. A primary impetus behind this standard effort (which was dubbed *VITAL*, for *VHDL Initiative Toward ASIC Libraries*) was to make it easier for ASIC vendors and others to generate timing models applicable to both VHDL and Verilog HDL. For this reason, the underlying data formats of IEEE 1076.4 and Verilog's SDF are quite similar.

Since Standard 1076.4 is only of interest to simulation model developers, we will cover it only briefly in this book. You should know, however, that ASIC and FPGA vendors are moving quickly toward timing model solutions built upon Standard 1076.4. As a result, while you may never need to describe a simulation model using Standard 1076.4, you can expect to use timing models provided to you by ASIC and FPGA vendors that are based on 1076.4.

How Is VHDL Used?

VHDL is a powerful language that can be used at many levels to describe an electronic system. Just as a software programming language, such as Pascal, C or C++, can be used to develop software at a high level, VHDL can be used to develop hardware using abstract, high-level design methods.

There are many points in the overall design process where VHDL can help.

For design specification ("specify")

VHDL can be used right up front, while you are still designing at a high-level, to capture the performance and interface requirements of each component in a large system. This is particularly useful for large projects involving many team members. Using a top-down approach to design, a system designer may define the interface to each component in the system, and describe the acceptance requirements of those components in the form of a high-level test bench. The interface definition (typically expressed as a VHDL entity declaration) and higher-level performance specification (perhaps written in the form of a test bench) can then be passed on to other team members for completion.

For design entry ("capture")

Design capture is that phase in which the details of a design are entered (captured) in a computer-based design system. In this phase, you may express your design (or portions of your design) as schematics (either board-level or purely functional) or using VHDL descriptions. If you are planning to use synthesis technology, you will want to write the VHDL portions of the design using a style of VHDL that is appropriate for synthesis.

For design simulation ("verify")

Once entered into a computer-based design system, you may wish to simulate the operation of your circuit to find out if it will meet the functional and timing requirements of your design specification. For a complete verification, you will probably want to perform this simulation at two (or more) key points in the design process.

Simulation can help identify errors and verify that the final chip will work in-circuit.

First, if you created one or more test benches as part of your design specification, you will use a simulator to apply the test bench to your design as it was originally written. This simulation, which will probably ignore issues such as the detailed timing characteristics of the design, will give you confidence that the design operates as expected. This first pass at simulation is called functional simulation, which will uncover most logical errors in the design.

After you have moved on to the implementation phase and synthesized your design into some lower-level, technology specific representation (such as an FPGA or ASIC netlist), you will probably want to simulate the design once again, using the detailed timing data generated as a result of that implementation. Simulation at this level can help identify errors related to actual timing delays and can also verify to a high degree of accuracy that the final chip will work in-circuit.

For design documentation ("formalize")

The structured programming features of VHDL, coupled with its configuration management features, make VHDL a natural form in which to document a large and complex circuit. The value of using a high-level language such as VHDL for design documentation is pointed out by the fact that the U.S. Department of Defense now requires VHDL as the standard format for communicating design requirements between government subcontractors.

As an alternative to schematics

Schematics have long been a part of electronic system design, and it is unlikely that they will become extinct anytime soon. Schematics have their advantages, particularly when used to depict circuitry in block diagram form. For this reason many VHDL design tools now offer the ability to combine schematic and VHDL representations in a design.

As an alternative to proprietary languages

VHDL is replacing proprietary languages throughout the digital design world.

If you have used programmable logic devices in the past, then you have probably already used some form of hardware description language (HDL). Proprietary languages such as PALASM, ABEL, CUPL and Altera's AHDL have been developed over the years by PLD device vendors and design tool suppliers, and they remain in widespread use today. In fact, there are probably more users of PLD-oriented proprietary languages in the word today than all other HDLs (including Verilog HDL and VHDL) combined.

The problem with proprietary languages is that they are non-portable, and your selection of design tools using these languages is quite limited. For these reasons, a standard language such as VHDL has a much broader appeal. We are now seeing VHDL (and its main rival, Verilog HDL) begin to replace proprietary languages throughout the digital design world.

Another key difference between VHDL and simpler PLD-oriented languages is that VHDL can be used to describe the test environment (in the form of a test bench) as well as the design itself.

The FPGA/ASIC Design Process

The following diagram shows a simplified design process including both synthesis and simulation, assuming that the target of the process is one or more programmable logic or ASIC chips. The key to understanding this process, and to understanding how best to use VHDL, is to remember the importance of test development. Test development should begin as soon as the general requirements of the system are known.

VHDL can be used for both design and test development.

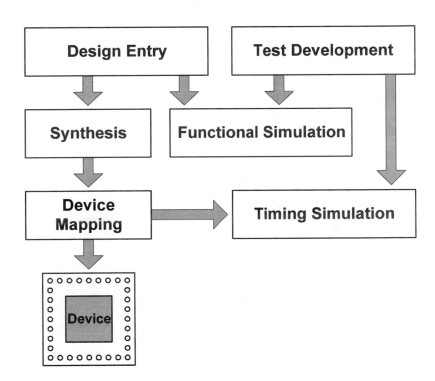

Where does VHDL fit in this diagram? VHDL (along with other forms of entry, such as schematics and block diagrams), will be used for design entry. After being captured into a design entry system using a text editor (or via a design entry tool that generates VHDL from higher-level graphical representations), the VHDL source code can be the input to simulation, allowing it to be functionally verified, or it can be passed directly to synthesis tools for implementation in a specified type of device.

On the test development side, VHDL test benches can be created that exercise the circuit to verify that it meets the functional and timing constraints of the specification. These test benches may be entered using a text editor, or they may be generated from other forms of test stimulus information, such as graphical waveforms.

For accurate timing simulation of a post-route layout, you will probably use a timing model generation program obtained from a device vendor or third party simulation model supplier. Model generation tools such as these typically generate timing-annotated VHDL source files that support very accurate system-level simulation.

When Should You Use VHDL?

Why choose to use VHDL for your design efforts? There are many likely reasons. If you ask most VHDL tool vendors this question, the first answer you will get is, "It will improve your productivity." But just what does this mean? Can you really expect to get your projects done faster using VHDL than by using your existing design methods?

The answer is yes, but probably not the first time you use it, and only if you apply VHDL in a structured manner. VHDL (like a structured software design language) is most beneficial when you use a structured, top-down approach to design. Real increases in productivity will come later, when you have climbed higher on the VHDL learning curve and have accumulated a library of reusable VHDL components.

Productivity increases will also occur when you begin to use VHDL to enhance communication between team members and when you take advantage of the more powerful tools for simulation and design verification that are available. In addition, VHDL allows you to design at a more abstract level. Instead of focusing on a gate-level implementation, you can address the behavioral function of the design.

VHDL makes it easy to build, use, and reuse libraries of circuit elements.

How will VHDL increase your productivity? By making it easy to build and use libraries of commonly-used VHDL modules. VHDL makes design reuse feel natural. As you discover the benefits of reusable code, you will soon find yourself thinking of ways to write your VHDL statements in ways that make them general purpose. Writing portable code will become an automatic reflex.

VHDL can greatly improve your chances of moving into more advanced tools and device targets.

Another important reason to use VHDL is the rapid pace of development in electronic design automation (EDA) tools and in target technologies. Using a standard language such as VHDL can greatly improve your chances of moving into more advanced tools (for example, from a basic low-cost simulator to a more advanced one) without having to re-enter your circuit descriptions. Your ability to retarget circuits to new types of device targets (for example, ASICs, FPGAs, and complex PLDs) will also be improved by using a standard design entry method.

When Should You Not Use VHDL?

To be fair, let's examine some of the most common reasons why you might not want to use VHDL as your primary method of design entry.

Steep learning curve?

Getting started is easy.

As described previously, there is a popular (and to some extent deserved) perception that VHDL is difficult to learn and use. While it is true that VHDL is a large and complex language, there is certainly no need to learn all of its features before beginning with your first few designs. Many of the advanced (and obscure) features of VHDL are useful only for simulation modeling or are intended for advanced configuration management. These parts of the language are rarely used when you are designing for synthesis. If you stick with well-established coding conventions such as those described in this book, then you should have little trouble getting started. As you gain confidence and experience in the language, you will naturally begin to explore some of the more advanced features.

Scarcity of low-cost tools?

VHDL-related design tools have historically carried a higher price tag than other alternatives, due in part to the fact that most of these tools were only available as smaller parts of large, workstation-based EDA tools. More recently, however, lower cost simulation and synthesis tools have appeared that make VHDL design tools more competitive.

Lack of technology-specific features?

VHDL's design management and modularity features allow you to accommodate different synthesis tools.

One of VHDL's major strengths — its general-purpose design — is also one of its primary weaknesses when compared to other existing tools, most notably PLD and FPGA design tools. While a PLD language such as PALASM, ABEL or CUPL provides specialized language features for accessing common device features (such as flip-flops and I/O buffers) and for specifying physical data (such as pin numbers), VHDL provides no such features. This means that synthesis tool developers must publish conventions (such as special attributes, comment fields or VHDL coding standards) to allow their users to pass such information into the synthesis tool. These conventions are rarely common between different tools, so users are faced with a choice: either write completely technology-independent VHDL code that may not be appropriate for the target device or technology, or write VHDL code that is specific to a particular vendor's tools.

This problem is actually not as bad as it sounds. The conventions published and used by VHDL synthesis vendors are designed in such a way that their use does not conflict with the use of similar features in alternative tools. Special attributes for pin numbers, for example, will be ignored by tools that do not have those attributes defined. If you write VHDL source code for use with more than one synthesis tool, you will probably use VHDL's design management and modularity features to isolate the technology-specific statements to easily-modified sections of your design.

Lack of VHDL applications expertise?

If you have made extensive use of PLDs and FPGAs, you may have become used to calling your local device vendor representative for help and advice. PLD and FPGA application engineers generally possess craniums full of useful knowledge about PLD and FPGA applications, and they are often well versed in the use of PLD and FPGA tools. Unfortunately, these same application engineers may not be completely familiar with VHDL, and they are probably not experts in using VHDL for PLD and FPGA applications.

PLD and FPGA vendors are adopting VHDL as a standard design entry method for their devices.

This situation is changing, however, as PLD and FPGA vendors adopt VHDL as a standard design entry method for their devices and begin offering VHDL-based design solutions to their customers. In addition, you can often find examples of VHDL source code or application notes at vendors' Web sites or ftp locations.

What About Verilog?

When it comes to hardware description languages, VHDL is not the only game in town. Verilog HDL has been around for many years as well, and it has a large number of fans who use it for both simulation and synthesis.

How does Verilog differ from VHDL? First, Verilog has some advantage in availability of simulation models. Soon after its introduction (by Gateway Design Automation in the mid-1980s), Verilog established itself as the de facto standard language for ASIC simulation libraries. The widespread popularity of Verilog HDL for this purpose had left VHDL somewhat behind in support for a wide range of ASIC devices. To solve this problem, the VITAL initiative (IEEE Standard 1076.4) enhances VHDL with a standard method of delay annotation that is similar to that used in Verilog. VITAL makes it easier for ASIC vendors and others to quickly create VHDL models from existing Verilog libraries.

Another feature that is defined in Verilog (in the Open Verilog International published standard) is a programming language interface, or PLI. The PLI makes it possible for simulation model writers to go outside of Verilog when necessary to create faster simulation models, or to create functions (using the C language) that would be difficult or inefficient to implement directly in Verilog.

The bottom line? Arguing about language features is pointless and best left to late night sessions at your favorite watering hole.

In VHDL's favor are its adoption as an IEEE standard (Standard 1076) and its higher-level design management features. These design management features, which include VHDL's configuration declaration and library features, make VHDL an ideal language for large project use.

In most respects, however, VHDL and Verilog are identical in function. The syntax is different (with Verilog looking very much like C, and VHDL looking more like Pascal or Ada), but basic concepts are the same. Both languages are easy to learn and hard to master. And once you have learned one of these languages, you will have no trouble transitioning to the other.

The bottom line is this: arguing about language features is pointless and best left to late night sessions at your favorite watering hole. You should choose a set of tools to help you do your job. Select these tools based on their features and their support for your specific application and technology. Don't choose a hardware description language based only on how it looks.

What About PLD Languages?

VHDL is a general-purpose language, while PLD-oriented languages are specialized.

Many new users of VHDL have previous experience with simpler PLD-oriented languages such as PALASM, ABEL or CUPL. How do these languages differ from VHDL? The most important thing to keep in mind about VHDL and these simpler languages is that VHDL was developed as a general-purpose simulation modeling language, while PLD languages were developed as specialized languages for capture and synthesis of relatively small digital circuits.

PLD languages, because of their modest roots, usually have simple feature sets and are easy to master. They include features (such as built-in register primitives, pin numbering and output polarity controls) that are specific to the PLD and FPGA devices they are intended to support. Because these languages have been designed specifically for synthesis, they do not include features (such as delay specifiers) that are not synthesizable.

There are more powerful ways of testing a design written in VHDL than one written in a PLD-oriented language. In a PLD-oriented language, you test a design by using a list of hard-coded test vectors. By contrast, you can test VHDL-based designs by writing general purpose programs known as test benches.

Perhaps just as important as the general features of PLD and FPGA design languages is their low cost: many of these tools are actually available free from vendors of PLDs and FPGAs, or they are sold at prices far below what you would expect to pay for VHDL- or Verilog-based synthesis tools.

17

Let's Get Started

Now that you have a good background in VHDL, it's time to have some *real* fun actually using the language. We'll begin with...

2. A First Look At VHDL

In this chapter, we start with some very simple examples...

To help put VHDL into a proper context and emphasize its use as a design entry language, this chapter presents several sample circuits and shows how they can be described for synthesis and testing.

In addition to the quick introduction to VHDL presented in this chapter, there are some very important concepts that will be introduced. Perhaps the most important concepts to understand in VHDL are those of *concurrency* and *hierarchy*. Since these concepts are so important (and may be new to you), we will introduce both concurrency and hierarchy in these initial examples. First, though, we will present a very simple example so you can see what constitutes the minimum VHDL source file.

As you look at these examples and read the information in this chapter, you will begin to understand some of the most important concepts of VHDL, and you will have a better understanding of how the more detailed features (covered later) can be used.

Simple Example: A Comparator

We'll start this chapter off by looking at a very simple combinational circuit: an 8-bit comparator. The circuit is shown below in block diagram form:

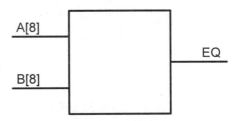

This comparator accepts two 8-bit inputs, compares them, and produces a 1-bit result (either 1, indicating a match, or 0, indicating a difference between the two input values).

Comparators and other typical combinational circuits can be described using simple VHDL language statements.

A comparator such as this is a combinational function constructed in circuitry from an arrangement of exclusive-OR gates or from some other lower-level structure, depending on the capabilities of the target technology. When described in VHDL, however, a comparator can be a simple language statement that makes use of VHDL's built-in relational operators.

Comparator Source File

VHDL includes many high-level language features that allow you to describe combinational logic circuits. The following VHDL source code uses a single concurrent assignment to describe the operation of our comparator:

```
-- Eight-bit comparator
--
entity compare is
    port( A, B: in bit_vector(0 to 7);
          EQ: out bit);
end compare;
```

```
architecture compare1 of compare is
begin

   EQ <= '1' when (A = B) else '0';

end compare1;
```

Note: In this and other VHDL source files listed in this book, VHDL keywords are highlighted in bold face type. In some VHDL books and software documents, keywords are highlighted by using upper case characters for keywords and lower case characters for identifiers. Some other books and manuals use lower case for keywords and upper case for identifiers.

Whatever forms you encounter or choose to use, keep in mind that VHDL itself is case-insensitive: keywords can be entered using either upper or lower case, and identifiers (such as signal and variable names) may be entered in either case as well, with no distinction being made between identifiers that are written in either case.

One more note: In the above context, the VHDL symbol <= is an assignment operator that assigns the value on its right to the signal on its left. Any text that follows "--" is a comment and is used for documentation only.

Now let's look more closely at this source file. Reading from the top, we see the following elements:

- An entity declaration that defines the inputs and outputs—the ports—of this circuit; and

- An architecture declaration that defines what the circuit actually does, using a single concurrent assignment.

Entities and Architectures

Every VHDL design description consists of at least one entity/architecture pair. (In VHDL jargon, this combination of an entity and its corresponding architecture is sometimes referred to as a *design entity*.) In a large design, you will typically write many entity/architecture pairs and connect them together to form a complete circuit.

A VHDL entity is analogous to a block symbol on a schematic.

An *entity declaration* describes the circuit as it appears from the "outside" - from the perspective of its input and output interfaces. If you are familiar with schematics, you might think of the entity declaration as being analogous to a block symbol on a schematic.

The second part of a minimal VHDL design description is the architecture declaration. Before simulation or synthesis can proceed, every referenced entity in a VHDL design description must be bound with a corresponding architecture. The architecture describes the actual function—or contents—of the entity to which it is bound. Using the schematic as a metaphor, you can think of the architecture as being roughly analogous to a lower-level schematic referenced by the higher-level functional block symbol.

Entity Declaration

The entity declaration provides a complete interface specification for a design unit.

An entity declaration provides the complete interface for a circuit. Using the information provided in an entity declaration (the names, data types and direction of each port), you have all the information you need to connect that portion of a circuit into other, higher-level circuits, or to develop input stimuli (in the form of a test bench) for verification purposes. The actual operation of the circuit, however, is not included in the entity declaration.

Let's take a closer look at the entity declaration for this simple design description:

```
entity compare is
    port( A, B: in bit_vector(0 to 7);
          EQ: out bit);
    end compare;
```

The entity declaration includes a name, **compare**, and a **port** statement defining all the inputs and outputs of the entity. The port list includes definitions of three ports: **A**, **B**, and **EQ**. Each of these three ports is given a direction (either **in**, **out** or **inout**), and a type (in this case either **bit_vector(0 to 7)**, which specifies an 8-bit array, or **bit**, which represents a single-bit value).

There are many different data types available in VHDL. We will cover these types later in this chapter, and in greater detail in Chapter 3, *Exploring Objects and Data Types*. To simplify things in this introductory circuit, we're going to stick with the simplest data types in VHDL, which are **bit** and **bit_vector**.

Architecture Declaration And Body

A VHDL architecture describes the actual contents of a design unit.

The second part of a minimal VHDL source file is the architecture declaration. Every entity declaration you reference in your design must be accompanied by at least one corresponding architecture (we'll discuss why you might have more than one architecture in a moment).

Here's the architecture declaration for the comparator circuit:

```
architecture compare1 of compare is
begin

    EQ <= '1' when (A = B) else '0';

end compare1;
```

The architecture declaration begins with a unique name, **compare1**, followed by the name of the entity to which the architecture is bound, in this case **compare**. Within the architecture declaration (between the **begin** and **end** keywords) is found the actual functional description of our comparator. There are many ways to describe combinational logic functions in VHDL; the method used in this simple design description is a type of concurrent statement known as a *condi-*

23

tional assignment. This assignment specifies that the value of the output (**EQ**) will be assigned a value of '1' when **A** and **B** are equal, and a value of '0' when they differ.

Concurrent assignments are the simplest form of VHDL statements.

This single concurrent assignment demonstrates the simplest form of a VHDL architecture. As you will see, there are many different types of concurrent statements available in VHDL, allowing you to describe very complex architectures. Hierarchy and subprogram features of the language allow you to include lower-level components, subroutines and functions in your architectures, and a powerful statement known as a *process* allows you to describe complex registered sequential logic as well.

Data Types

VHDL's high-level data types allow data to be represented in much the same way as in high-level programming languages.

Like a high-level software programming language, VHDL allows data to be represented in terms of high-level *data types.* A data type is an abstract representation of stored data, such as you might encounter in software languages. These data types might represent individual wires in a circuit, or they might represent collections of wires.

The preceding description of the comparator circuit used the data types **bit** and **bit_vector** for its inputs and outputs. The **bit** data type has only two possible values: '1' or '0'. (A **bit_vector** is simply an array of **bits.**) Every data type in VHDL has a defined set of values, and a defined set of valid operations. Type checking is strict, so it is not possible, for example, to directly assign the value of an **integer** data type to a **bit_vector** data type. (There are ways to get around this restriction, using what are called *type conversion functions.* These are discussed in Chapter 3, *Exploring Objects and Data Types.*)

The chart of Figure 2-1 summarizes the fundamental data types available in VHDL. Chapter 4, *Using Standard Logic,* describes additional standard data types available in VHDL.

Data Type	Values	Example
Bit	'1', '0'	Q <= '1';
Bit_vector	(array of bits)	DataOut <= "00010101";
Boolean	True, False	EQ <= True;
Integer	-2, -1, 0, 1, 2, 3, 4 . . .	Count <= Count + 2;
Real	1.0, -1.0E5	V1 = V2 / 5.3
Time	1 ua, 7 ns, 100 ps	Q <= '1' after 6 ns;
Character	'a', 'b', '2, '$', etc.	CharData <= 'X';
String	(Array of characters)	Msg <= "MEM: " & Addr

Figure 2-1: VHDL's built-in data types allow you to represent many types of data.

Note:
The VHDL symbol <= is an assignment operator that assigns the value(s) on its right to the variable on its left.

Design Units

Design units are a concept unique to VHDL that provide advanced configuration management capabilities.

One concept unique to VHDL (when compared to software programming languages and to its main rival, Verilog) is the concept of a design unit. Design units in VHDL (which may also be referred to as library units) are segments of VHDL code that can be compiled separately and stored in a library.

You have been introduced to two design units already: the entity and the architecture. There are actually five types of design units in VHDL; entities, architectures, packages, package bodies, and configurations. Entities and architectures are the only two design units that you must have in any VHDL design description. Packages and configurations are optional.

Figure 2-2 illustrates the relationship between these five design units, which are each described in the following sections.

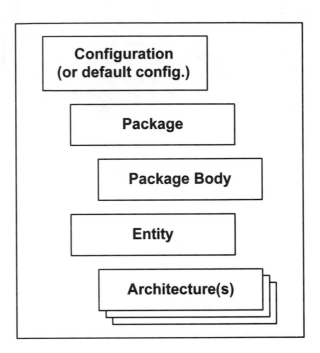

Figure 2-2: Design units are parts of a VHDL description that can be compiled (analyzed) into a library.

Entities

The minimum VHDL design description must include at least one entity and a corresponding architecture.

A VHDL *entity* is a statement (indicated by the **entity** keyword) that defines the external specification of a circuit or sub-circuit. The minimum VHDL design description must include at least one entity and one corresponding architecture.

When you write an entity declaration, you must provide a unique name for that entity and a port list defining the input and output ports of the circuit. Each port in the port list must be given a name, direction (or *mode*, in VHDL jargon) and a type. Optionally, you may also include a special type of parameter list (called a *generic list*, see Chapter 8) that allows you to pass additional information into an entity.

An example of an entity declaration is given below:

```
entity fulladder is
    port (X: in bit;
          Y: in bit;
          Cin: in bit;
```

```
            Cout: out bit;
            Sum: out bit);
   end fulladder;
```

Architectures

A VHDL architecture declaration is a statement (beginning with the **architecture** keyword) that describes the underlying function and/or structure of a circuit. Each architecture in your design must be associated (or *bound*) by name with one entity in the design.

VHDL allows you to create more than one alternative architecture for an entity.

VHDL allows you to create more than one alternate architecture for each entity. This feature is particularly useful for simulation and for project team environments in which the design of the system interfaces (expressed as entities) is performed by a different engineer than the lower-level architectural description of each component circuit, or when you simply want to experiment with different methods of description.

An architecture declaration consists of zero or more declarations (of items such as intermediate signals, components that will be referenced in the architecture, local functions and procedures, and constants) followed by a **begin** statement, a series of concurrent statements, and an **end** statement, as illustrated by the following example:

```
architecture concurrent of fulladder is
begin
   Sum <= X xor Y xor Cin;
   Cout <= (X and Y) or (X and Cin) or (Y and Cin);
end concurrent;
```

Packages and Package Bodies

A VHDL package declaration is identified by the **package** keyword, and is used to collect commonly-used declarations for use globally among different design units. You can think of a package as a common storage area, one used to store such things as type declarations, constants, and global subprograms. Items defined within a package can be made visible to any other design unit in the complete VHDL design, and they can be compiled into libraries for later re-use.

A package can consist of two basic parts: a package declaration and an optional package body.

A package can consist of two basic parts: a package declaration and an optional package body. Package declarations can contain the following types of statements, all of which are described in later chapters:

- Type and subtype declarations

- Constant declarations

- Global signal declarations

- Function and procedure declarations

- Attribute specifications

- File declarations

- Component declarations

- Alias declarations

- Disconnect specifications

- Use clauses

Items appearing within a package declaration can be made visible to other design units through the use of a **use** statement, as we will see.

If the package contains declarations of subprograms (functions or procedures) or defines one or more deferred constants (constants whose value is not immediately given), then a *package body* is required in addition to the package declaration. A package body (which is specified using the **package body** keyword combination) must have the same name as its corresponding package declaration, but it can be located anywhere in the design, in the same or a different source file.

The relationship between a package and package body is somewhat akin to the relationship between an entity and its corresponding architecture.

The relationship between a package and package body is somewhat akin to the relationship between an entity and its corresponding architecture. (There may be only one package body written for each package declaration, however.) While the package declaration provides the information needed to use the items defined within it (the parameter list for a global procedure, or the name of a defined type or subtype), the actual behavior of such things as procedures and functions must be specified within package bodies.

An example of a package is given below:

```
package conversion is
    function to_vector (size: integer; num: integer) return std_logic_vector;
end conversion;

package body conversion is
    function to_vector(size: integer; num: integer) return std_logic_vector is
        variable ret: std_logic_vector (1 to size);
        variable a: integer;
    begin
        a := num;
        for i in size downto 1 loop
            if ((a mod 2) = 1) then
                ret(i) := '1';
            else
                ret(i) := '0';
            end if;
            a := a / 2;
        end loop;
        return ret;
    end to_vector;
end conversion;
```

Examples of global procedures and functions can be found in Chapter 7, *Creating Modular Designs*.

Configurations

The final type of design unit available in VHDL is called a configuration declaration. You can think of a configuration declaration as being roughly analogous to a parts list for your design. A configuration declaration (identified with the **con-**

figuration keyword) specifies which architectures are to be bound to which entities, and it allows you to change how components are connected in your design description at the time of simulation. (Configurations are not generally used for synthesis, and may not be supported in the synthesis tool that you will use.)

Configuration declarations are always optional, no matter how complex a design description you create.

Configuration declarations are always optional, no matter how complex a design description you create. In the absence of a configuration declaration, the VHDL standard specifies a set of rules that provide you with a default configuration. For example, in the case where you have provided more than one architecture for an entity, the last architecture compiled will take precedence and will be bound to the entity.

Configurations are discussed at length in Chapter 8, *Partitioning Your Design*. An example is given below:

```
configuration this_build of rcomp is
    for structure
        for COMP1: compare use entity work.compare(compare1);
        for ROT1: rotate use entity work.rotate(rotate1);
    end for;
end this_build;
```

Structure Of A Small Design

Figure 2-3 illustrates one possible way to enter a small design description (perhaps one intended only for synthesis), in terms of its design unit relationships and actual source files:

For your first design efforts using VHDL, you may create design descriptions of similar complexity. To start with, you might create a single VHDL source file (using the text editor of your choice) and write a top-level entity declaration defining the input and output (I/O) of your circuit. You might then write an architecture corresponding to that entity describing the internal function or structure of the circuit.

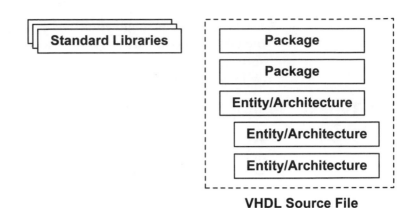

VHDL Source File

Figure 2-3: A very simple VHDL description might consist of just one source file containing a small number of entity/ architecture pairs.

If your project is more complex, you may write additional entity/architecture pairs, then connect these lower-level design units into your top-level architecture using VHDL's component instantiation features. (Component instantiation refers to VHDL's ability to connect a component reference in one entity with its declaration elsewhere.) Finally, you may choose to write one or more packages containing commonly-used items such as subprograms, constants or type declarations.

In addition to the information you supply in your primary source file(s), you will probably also make use of standard libraries (specifically the built-in library defined in IEEE Standard 1076 and the IEEE 1164 standard library). These libraries will have been provided to you by your simulation and/or synthesis tool vendors, and they will be referenced from within your VHDL source file as needed.

It is possible to describe an entire design in a single VHDL source file, but it is not a recommended method of design...

Although a design description such as this is easy to comprehend, easy to manage and easy to write, it is not a recommended method of design. First, because your synthesis software will not be capable of distinguishing between those parts of your description that are logic to be synthesized, and those that are for verification purposes only (the test bench), you will not be able to process the same design file through both simulation and synthesis. Also, you will find that it is

31

much easier to manage your design in simulation if you include only one entity and corresponding architecture in each source file.

More Typical Design Description

When creating a more complex design description, perhaps one intended for use in a team environment, you may choose to make use of VHDL's many features for design management and take a more structured approach to your design. Figure 2-4 illustrates one possible structure for such a project:

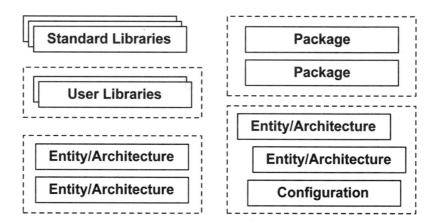

Figure 2-4: A more typical VHDL description consists of multiple VHDL source files.

In this design description, the project has been split into multiple source files as indicated by the dashed lines. Commonly-used packages have been placed into a user-defined library, and a configuration declaration has been provided that ties the whole project together.

Project or company-wide libraries of functions or lower-level components can dramatically improve design re-use...

The use of project or company-wide libraries of functions or lower-level components can dramatically improve design re-use, which can in turn lead to increased design productivity. As we will see in a later chapter, VHDL provides many features to help in such situations.

Levels of Abstraction (Styles)

VHDL supports many possible styles of design description. These styles differ primarily in how closely they relate to the underlying hardware. When we speak of the different styles of VHDL, we are really talking about the differing levels of abstraction possible using the language—behavior, dataflow, and structure—as shown in Figure 2-5.

Figure 2-5: Level of Abstraction refers to how far your design description is from an actual hardware realization.

Behavior	Performance Specifications Test Benches Sequential Descriptions State Machines
Dataflow	Register Transfers Selected Assignments Arithmetic Operations Boolean Equations
Structure	Hierarchy Physical Information

Level of Abstraction

The VHDL design process involves a top-down refinement of a design concept into a verifiable and synthesizable design description.

The chart of Figure 2-5 maps the various points in a top-down design process to the three general levels of abstraction. Starting at the top, suppose the performance specifications for a given project are: "the compressed data coming out of the DSP chip needs to be analyzed and stored within 70 nanoseconds of the strobe signal being asserted..." This human language specification must be refined into a description that can actually be simulated. A test bench written in combination with a sequential description is one such expression of the design. These are all points in the **behavior** level of abstraction.

After this initial simulation, the design must be further refined until the description is something a VHDL synthesis tool can digest. Synthesis is a process of translating an abstract concept into a less-abstract form. The highest level of abstraction accepted by today's synthesis tools is the **dataflow** level.

The **structure** level of abstraction comes into play when little chunks of circuitry are to be connected together to form bigger circuits. (If the little chunks being connected are actually quite large chunks, then the result is what we commonly call a block diagram.) Physical information is the most basic level of all and is outside the scope of VHDL. This level involves actually specifying the interconnects of transistors on a chip, placing and routing macrocells within a gate array or FPGA, etc.

Note:

In some formal discussions of synthesis, four levels of abstraction are described; behavior, RTL, gate-level and layout. It is our view that the three levels of abstraction presented here provide the most useful distinctions for today's synthesis user.

Any given design can be represented in VHDL in a variety of different ways, using different levels of abstraction and corresponding language features.

As an example of these three levels of abstraction, it is possible to describe a complex controller circuit in a number of ways. At the lowest level of abstraction (the structural level), we could use VHDL's hierarchy features to connect a sequence of predefined logic gates and flip-flips to form the complete circuit. To describe this same circuit at a dataflow level of abstraction, we might describe the combinational logic portion of the controller (its input decoding and transition logic) using higher-level Boolean logic functions and then feed the output of that logic into a set of registers that match the registers available in some target technology. At the behavioral level of abstraction, we might ignore the target technology (and the requirements of synthesis tools) entirely and instead describe how the controller operates over time in response to various types of stimulus.

Behavior

The highest level of abstraction supported in VHDL is called the *behavioral* level of abstraction. When creating a behavioral description of a circuit, you will describe your circuit in terms of its operation *over time*. The concept of time is the critical distinction between behavioral descriptions of circuits and lower-level descriptions (specifically descriptions created at the dataflow level of abstraction).

Examples of behavioral forms of representation might include state diagrams, timing diagrams and algorithmic descriptions (Figure 2-6).

Figure 2-6: Behavioral descriptions (such as this state diagram with accompanying source code) define the operation of a circuit over time.

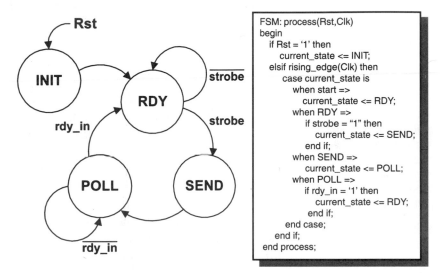

```
FSM: process(Rst,Clk)
begin
    if Rst = '1' then
        current_state <= INIT;
    elsif rising_edge(Clk) then
        case current_state is
            when start =>
                current_state <= RDY;
            when RDY =>
                if strobe = "1" then
                    current_state <= SEND;
                end if;
            when SEND =>
                current_state <= POLL;
            when POLL =>
                if rdy_in = '1' then
                    current_state <= RDY;
                end if;
        end case;
    end if;
end process;
```

Using a behavioral level of abstraction, you can describe the operation of a sequential circuit in such a way that registers are implied.

In a behavioral description, the concept of time may be expressed precisely, with actual delays between related events (such as the propagation delays within gates and on wires), or it may simply be an ordering of operations that are expressed sequentially (such as in a functional description of a flip-flop). When you are writing VHDL for input to synthesis tools, you may use behavioral statements in VHDL to imply that there are registers in your circuit. It is unlikely, however, that your synthesis tool will be capable of creating precisely the same behavior in actual circuitry as you have defined in the language. (Synthesis tools today ignore detailed timing specifications, leaving the actual timing results at the mercy of the target device technology.) It is also unlikely that your synthesis tool will be capable of accepting and processing a very wide range of behavioral description styles.

Writing behavior-level VHDL is very much like writing one or more small software programs.

If you are familiar with software programming, writing behavior-level VHDL will not seem like anything new. Just like a programming language, you will be writing one or more small programs that operate sequentially and communicate with one another through their interfaces. The only difference between behavior-level VHDL and a software programming language is the underlying execution platform: in the case of software, it is some operating system running on a CPU; in the case of VHDL, it is the simulator and/or the synthesized hardware.

Dataflow

In the dataflow level of abstraction you describe how information is passed between registers in your circuit.

In the dataflow level of abstraction, you describe your circuit in terms of how data moves through the system. At the heart of most digital systems today are registers, so in the dataflow level of abstraction you describe how information is passed between registers in the circuit. You will probably describe the combinational logic portion of your circuit at a relatively high level (and let a synthesis tool figure out the detailed implementation in logic gates), but you will likely be quite specific about the placement and operation of registers in the complete circuit.

An example of the dataflow level of abstraction is shown in Figure 2-7. This circuit (a 20-bit loadable counter) would be very tedious to describe using structural techniques.

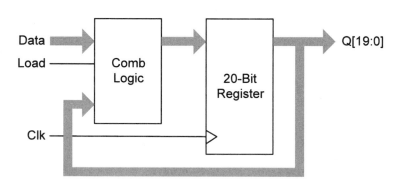

Figure 2-7: At the dataflow level of abstraction, a large counter is described in terms of registers and combinational logic.

The dataflow level of abstraction allows the drudgery of combinational logic to be simplified...

The dataflow level of abstraction is often called *register transfer logic*, or RTL. This level of abstraction is an intermediate level that allows the drudgery of combinational logic to be simplified (and, presumably, taken care of by logic synthesis tools) while the more important parts of the circuit, the registers, are more completely specified.

There are some drawbacks to using a dataflow method of design in VHDL. First, there are no built-in registers in VHDL; the language was designed to be general-purpose, and the emphasis was placed by VHDL's designers on its behavioral aspects. If you are going to write VHDL at the dataflow level of abstraction, you must first create (or obtain) behavioral descriptions of the register elements you will be using in your design. These elements must be provided in the form of components (using VHDL's hierarchy features) or in the form of subprograms (functions or procedures).

But for hardware designers, it can be difficult to relate the sequential descriptions and operation of behavioral VHDL with the hardware being described (or *modeled*). For this reason, many VHDL users, particularly those who are using VHDL as an input to synthesis, prefer to stick with levels of abstraction that are easier to relate to actual hardware devices (such as logic gates and flip-flops). These users are often more comfortable using the dataflow level of abstraction.

Structure

A structure-level description defines the circuit in terms of a collection of components.

The third level of abstraction, *structure*, is used to describe a circuit in terms of its components. Structure can be used to create a very low-level description of a circuit (such as a transistor-level description) or a very high-level description (such as a block diagram).

In a gate-level description of a circuit, for example, components such as basic logic gates and flip-flops might be connected in some logical structure to create the circuit. This is what is often called a *netlist*. For a higher-level circuit—one in which the components being connected are larger functional blocks—structure might simply be used to segment the design description into manageable parts.

Structure-level VHDL features are very useful for managing complexity.

Structure-level VHDL features, such as components and configurations, are very useful for managing complexity. The use of components can dramatically improve your ability to re-use elements of your designs, and they can make it possible to work using a *top-down* design approach.

To give an example of how a structural description of a circuit relates to higher levels of abstraction, consider the design of a simple 5-bit counter. To describe such a counter using traditional design methods, we might immediately draw a picture of five T flip-flops with some simple decode logic as shown in Figure 2-8:

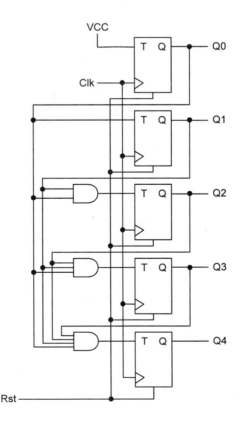

Figure 2-8: At the structure level of abstraction, you connect together lower-level components, such as gates and flip-flops, to create a larger system.

The following VHDL design description represents the preceding schematic in the form of a netlist of connected components:

```
entity andgate is
    port(A,B,C,D: in bit := '1'; Y: out bit);
end andgate;

architecture gate of andgate is
begin
    Y <= A and B and C and D;
end gate;

entity tff is
    port(Rst,Clk,T: in bit; Q: out bit);
end tff;


architecture behavior of tff is
begin
    process(Rst,Clk)
        variable Qtmp: bit;
    begin
        if (Rst = '1') then
            Qtmp := '0';
        elsif Clk = '1' and Clk'event then
            if T = '1' then
                Qtmp := not Qtmp;
            end if;
        end if;
        Q <= Qtmp;
    end process;
end behavior;

entity TCOUNT is
    port (Rst,Clk: in bit;
        Count: out bit_vector(4 downto 0));
end TCOUNT;

architecture STRUCTURE of TCOUNT is
    component tff
        port(Rst,Clk,T: in bit; Q: out bit);
    end component;
    component andgate
        port(A,B,C,D: in bit := '1'; Y: out bit);
    end component;
```

```
        constant VCC: bit := '1';
        signal T,Q: bit_vector(4 downto 0);

    begin
        T(0) <= VCC;
        T0: tff port map (Rst=>Rst, Clk=>Clk, T=>T(0), Q=>Q(0));
        T(1) <= Q(0);
        T1: tff port map (Rst=>Rst, Clk=>Clk, T=>T(1), Q=>Q(1));
        A1: andgate port map(A=>Q(0), B=>Q(1), Y=>T(2));
        T2: tff port map (Rst=>Rst, Clk=>Clk, T=>T(2), Q=>Q(2));
        A2: andgate port map(A=>Q(0), B=>Q(1), C=>Q(2), Y=>T(3));
        T3: tff port map (Rst=>Rst, Clk=>Clk, T=>T(3), Q=>Q(3));
        A3: andgate port map(A=>Q(0), B=>Q(1), C=>Q(2), D=>Q(3), Y=>T(4));
        T4: tff port map (Rst=>Rst, Clk=>Clk, T=>T(4), Q=>Q(4));

        Count <= Q;

    end STRUCTURE;
```

This structural representation seems a straightforward way to describe a 5-bit counter, and it is certainly easy to relate to hardware since just about any imaginable implementation technology will have the features necessary to implement the circuit. For larger circuits, however, such descriptions quickly become impractical.

Sample Circuit

To help demonstrate some of the important concepts we have covered in the first half of this chapter, we will present a very simple circuit and show how the function of this circuit can be described in VHDL. The design descriptions we will show are intended for synthesis and therefore do not include timing specifications or other information not directly applicable to today's synthesis tools.

We will present a simple circuit that includes both combinational and registered logic elements.

This sample circuit combines the comparator circuit presented earlier with a simple 8-bit loadable shift register. The shift register will allow us to examine in detail how higher-level VHDL descriptions can be written for synthesis of both combinational and registered logic.

The two subcircuits (the shifter and comparator) will be connected using VHDL's hierarchy features and will demonstrate the third level of abstraction: *structure*. The complete circuit is shown in Figure 2-9.

Figure 2-9: Our sample circuit will combine behavior, dataflow, and structure levels of abstraction.

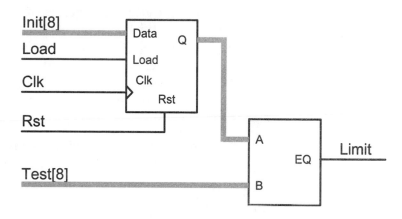

Many of the concepts of VHDL are familiar to users of schematic hierarchy.

This diagram has been intentionally drawn to look like a hierarchical schematic with each of the lower-level circuits represented as blocks. In fact, many of the concepts we will cover during the development of this circuit are the same concepts familiar to users of schematic hierarchy. These concepts include the ideas of component instantiation, mapping of ports, and design partitioning.

In a more structured project environment, you would probably enter a circuit such as this by first defining the interface requirements of each block, then describing the overall design of the circuit as a collection of blocks connected together through hierarchy at the top level. Later, after the system interfaces had been designed, you would proceed down the hierarchy (using a top-down approach to design) and fill in the details of each subcircuit.

In this example, however, we will begin by describing each of the lower-level blocks first and then connect them to form the complete circuit.

Comparator

Using standard logic data types allows circuit elements from different sources to be easily combined and re-used, and gives you more detailed and precise simulation.

The comparator portion of the design will be identical to the simple 8-bit comparator we have already seen. The only difference is that we will use the IEEE 1164 standard logic data types (**std_logic** and **std_logic_vector**) rather than the **bit** and **bit_vector** data types used previously. Using standard logic data types for all system interfaces is highly recommended, as it allows circuit elements from different sources to be easily combined. It also provides you the opportunity to perform more detailed and precise simulation than would otherwise be possible.

The updated comparator design description, using the IEEE 1164 standard logic data types, is shown below:

```
-------------------------------
-- Eight-bit comparator
--
library ieee;
use ieee.std_logic_1164.all;
entity compare is
    port (A, B: in std_logic_vector(0 to 7);
          EQ: out std_logic);
end compare;

architecture compare1 of compare is
begin
    EQ <= '1' when (A = B) else '0';
end compare1;
```

Let's take a closer look at this simple VHDL design description. Reading from the top of the source file, we see:

- a comment field, indicated by the leading double-dash symbol ("--"). VHDL allows comments to be embedded anywhere in your source file, provided they are prefaced by the two hyphen characters as shown. Comments in VHDL extend from the double-dash symbol to the end of the current line. There is no block comment facility in VHDL.

- a **library** statement that causes the named library IEEE to be loaded into the current compile session. When you use VHDL libraries, it is recommended that you include your library statements once at the beginning of the source file, before any **use** clauses or other VHDL statements.

- a **use** clause, specifying which items from the IEEE library are to be made available for the subsequent design unit (the entity and its corresponding architecture). The general form of a **use** statement includes three fields delimited by a period: the library name (in this case **ieee**), a design unit within the library (normally a package, in this case named **std_logic_1164**), and the specific item within that design unit (or, as in this case, the special keyword **all**, which means everything) to be made visible.

- an entity declaration describing the interface to the comparator. Note that we have now specified **std_logic** and **std_logic_vector**, which are standard data types provided in the IEEE 1164 standard and in the associated IEEE library.

- an architecture declaration describing the actual function of the comparator circuit.

Conditional Signal Assignment

The function of the comparator is defined using a simple concurrent assignment to port **EQ**. The type of statement used in the assignment to **EQ** is called a *conditional signal assignment*.

Conditional signal assignments allow complex combinational logic to be described.

Conditional signal assignments make use of the **when-else** language feature and allow complex conditional logic to be described. The following VHDL description of a multiplexer (diagrammed in Figure 2-10) makes the use of the conditional signal assignment more clear:

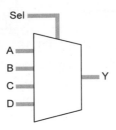

Figure 2-10. A conditional signal assignment describes two or more alternate conditions, as in this 4-input multiplexer.

```
architecture mux1 of mux is
begin
    Y <=  A when (Sel = "00") else
          B when (Sel = "01") else
          C when (Sel = "10") else
          D when (Sel = "11");
end mux1;
```

Selected Signal Assignment

Another form of signal assignment can be used as an alternative to the conditional signal assignment. The *selected signal assignment* has the following general form (again, using a multiplexer as an example):

The selected signal assignment is similar to the conditional signal assignment, but it does not imply a priority for the conditions.

```
architecture mux2 of mux is
begin
    with Sel select
        Y <=  A when "00",
              B when "01",
              C when "10",
              D when "11";
end mux2;
```

Choosing between a conditional or selected signal assignment for circuits such as this is largely a matter of taste. For most designs, there is no difference in the results obtained with either type of assignment statement. For some circuits, however, the conditional signal assignment can imply priorities that result in additional logic being required. This difference is discussed in detail in Chapter 5, *Understanding Concurrent Statements*.

Note:

In the 1993 version of the VHDL language standard, there is one important difference between the conditional signal assignment and a selected signal assignment that may cause some confusion. In Standard 1076-1993, a conditional assignment statement does not have to include a final, terminating else clause, and the conditions specified do not have to be all-inclusive. When one or more possible input conditions are missing from a conditional assignment, the default behavior under those conditions is for the target signal to hold its value. This means that the conditional assignment statement can be used to effectively create a latch. Some synthesis tools support this use of a conditional signal assignment, while others do not. The best advice: do not write incomplete conditional signal assignments.

Shifter (Entity)

The second and most complex part of this design is the 8-bit shifter. This circuit (diagrammed below in Figure 2-11) accepts 8-bit input data, loads this data into a register and, when the **load** input signal is low, rotates this data by one bit with each rising edge clock signal. The shift register is provided with an asynchronous reset, and the data stored in the register are accessible via the output signal **Q**.

Figure 2-11: The 8-bit shifter accepts an 8-bit value (Data) when the Load input is asserted. Otherwise, it rotates the contents of its registers by one bit. The shifter also includes an asynchronous reset.

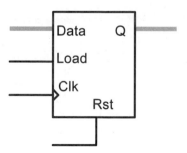

They are many ways to describe a shifter such as this in VHDL. If you are going to use synthesis tools to process the design description into an actual device technology, however, you must restrict yourself to well established synthesis conventions when entering the design description. We will examine two of these conventions when entering this design.

Using a Process

A process can be used to describe the behavior of a circuit over time.

The first design description that we will write for this shifter is a description that uses a VHDL **process** statement to describe the behavior of the entire circuit over time. The process is a behavioral language feature, but its use for synthesis is quite constrained, as we will see in this and later chapters.

The VHDL source code for the shifter is shown below:

```
----------------------------------------------------------
-- Eight-bit  shifter
--
library ieee;
use ieee.std_logic_1164.all;
entity rotate is
    port( Clk, Rst, Load: in std_logic;
            Data: in std_logic_vector(0 to 7);
            Q: out std_logic_vector(0 to 7));
end rotate;


architecture rotate1 of rotate is
    signal Qreg: std_logic_vector(0 to 7);
begin
    reg: process(Rst,Clk)
    begin
        if Rst = '1' then   -- Async reset
            Qreg <= "00000000";
        elsif (Clk = '1' and Clk'event) then
            if (Load = '1') then
                Qreg <= Data;
            else
                Qreg <= Qreg(1 to 7) & Qreg(0);
            end if;
        end if;
    end process;
    Q <= Qreg;
end rotate1;
```

Let's look closely at this source file. Reading from the top, we see:

- a comment field, as described previously.

- **library** and **use** statements, allowing us to use the IEEE 1164 standard logic data types.

- an entity declaration defining the interface to the circuit. Note that the direction (mode) of **Q** is written as **out,** indicating that we will not attempt to read its value within this design unit. (If **Q** was to be used directly as the register, rather than introducing an intermediate signal **Qreg**, it would need to be of mode **inout** or **buffer.**)

- an architecture declaration, consisting of a single **process** statement that defines the operation of the shifter over time in response to events appearing on the clock (**Clk**) and asynchronous reset (**Rst**).

Process Statement

The process statement in VHDL is the primary means by which sequential operations (such as registered circuits) can be described. For use in describing registered circuits, the most common form of a process statement is:

*A **process** statement allows sequential statements to be described, which can imply operation over time.*

```
architecture arch_name of ent_name is
begin
    process_name: process(sensitivity_list)
        local_declaration;
        local_declaration;
        . . .
    begin
        sequential statement;
        sequential statement;
        sequential statement;
        .
        .
        .
    end process;
end arch_name;
```

A process statement consists of the following items:

- An optional process name (an identifier followed by a colon character).

- The **process** keyword.

- An optional sensitivity list, indicating which signals result in the process being executed when there is some event detected. (The sensitivity list is required if the process does not include one or more **wait** statements to suspend its execution at certain points. We will look at examples that do not use a sensitivity list later on in this chapter).

- An optional declarations section, allowing local objects and subprograms to be defined.

- A **begin** keyword.

- A sequence of statements to be executed when the process runs.

- an **end process** statement.

A process describes the sequential execution of statements that depend on one or more events occurring.

The easiest way to think of a VHDL process such as this is to relate it to a small software program that executes (in simulation) any time there is an event on one of the process inputs, as specified in the sensitivity list. A process describes the sequential execution of statements that are dependent on one or more events occurring. A flip-flop is a perfect example of such a situation; it remains idle, not changing state, until there is a significant event (either a rising edge on the clock input or an asynchronous reset event) that causes it to operate and potentially change its state.

Although there is a definite order of operations within a process (from top to bottom), you can think of a process as executing in zero time. This means that (a) a process can be used to describe circuits functionally, without regard to their actual timing, and (b) multiple processes can be "executed" in parallel with little or no concern for which processes complete their operations first. (There are certain caveats to this behavior of VHDL processes. These caveats will be described in detail in Chapter 6, *Understanding Sequential Statements*.)

Let's see how the process for our shifter operates. For your reference, the process is shown below:

```
reg: process(Rst,Clk)
begin
  if Rst = '1' then   -- Async reset
    Qreg <= "00000000";
  elsif (Clk = '1' and Clk'event) then
    if (Load = '1') then
      Qreg <= Data;
    else
      Qreg <= Qreg(1 to 7) & Qreg(0);
    end if;
  end if;
end process;
```

In the absence of an event on one or more of the signals in the sensitivity list, the process remains suspended and the signals assigned within the process hold their values.

As written, the process is dependent on (or *sensitive* to) the asynchronous inputs **Rst** and **Clk**. These are the only signals that can have events directly affecting the operation of the circuit; in the absence of any event on either of these signals, the circuit described by the process will simply hold its current value (that is, the process will remain suspended).

Now let's examine what happens when an event occurs on either one of these asynchronous inputs. First, consider what happens when the input **Rst** has an event in which it transitions to a high state (represented by the **std_ulogic** value of '1'). In this case, the process will begin execution and the first **if** statement will be evaluated. Because the event was a transition to '1', the simulator will see that the specified condition (**Rst = '1'**) is true and the assignment of signal **Qreg** to the reset value of "00000000" will be performed. The remaining statements of the **if-then-elsif** expression (those that are dependent on the **elsif** condition) will be ignored. (The assignment statement immediately following the process, the assignment of output signal **Q** to the value of **Qreg**, is not subject to the **if-then-elsif** expression of the process or its sensitivity list, and is therefore valid at all times.) Finally, the process suspends, all signals that were assigned values in the process (in this case **Qreg**) are updated, and the process waits for another event on **Clk** or **Rst**.

A process can be somewhat confusing, but it is important to understand how signals within the process are updated over time.

What about the case in which there is an event on **Clk**? In this case, the process will again execute, and the **if-then-elsif** expressions will be evaluated, in turn, until a valid condition is encountered. If the **Rst** input continues to have a high value (a value of **'1'**), then the simulator will evaluate the first **if** test as true, and the reset condition will take priority. If, however, the **Rst** input is not a value of **'1'**, then the next expression (**Clk = '1' and Clk'event**) will be evaluated. This expression is the most commonly-used convention for detecting clock edges in VHDL. To detect a rising edge clock, we write the expression **Clk = '1'** in the conditional expression, just as we did when detecting a reset condition. For this circuit, however, the expression **Clk = '1'** would not be specific enough, since the process may have begun execution as the result of an event on **Rst** that did not result in **Rst** transitioning to a **'1'**. To ensure that the event we are responding to is actually an event on **Clk**, we use the built-in VHDL attribute **'event** to check if **Clk** was that signal triggering the process execution.

If the event that triggered the process execution was actually a rising edge on **Clk**, then the simulator will go on to check the remaining **if-then** logic to determine which assignment statement is to be executed. If **Load** is determined to be **'1'**, then the first assignment statement is executed and the data is loaded from input **Data** to the registers. If **Load** is not **'1'**, then the data in the registers is shifted, as specified, using the bit slice and concatenation operations available in the language.

Confusing? Perhaps; but if you simply use the style just presented as a template for describing registered logic and don't worry too much about the details of how it is executed during simulation, you should be in pretty good shape. Just keep in mind that every assignment to a signal you make that is dependent on a **Clk = '1' and Clk'event** expression will result in at least one register when synthesized. (More detailed discussions of how flip-flops and latches are generated from synthesis tools can be found in Chapter 6, *Understanding Sequential Statements*.)

Process Statements Without Sensitivity Lists

VHDL process statements have two primary forms. The first form uses the sensitivity list described in the previous section and executes during simulation whenever there is an event on any signal in the sensitivity list. This first form of a process statement is the most common (and recommended) method for describing registered sequential logic for the purposes of synthesis.

There is another form of process statement that is useful for other applications, however. This form does not include a sensitivity list. Instead, it includes one or more statements that suspend the execution of the process until some condition has been met. The best example of such a process is a test bench, in which a sequence of test inputs are applied over time, with a predefined time value (or external triggering event) defined as the condition for re-activation of the process. The general form of such a process is:

*A process must include either a sensitivity list, or one or more **wait** statements.*

```
architecture arch_name of ent_name is
begin
  process_name: process
    local_declaration;
    local_declaration;
    . . .
  begin
    sequential statement;
    sequential statement;
    wait until (condition);
    sequential statement;
    . . .
    wait for (time);
    . . .
  end process;
end arch;
```

51

Examples of this form of process will be examined later in this chapter and in Chapter 9, *Writing Test Benches*.

VHDL requires that all processes include either a sensitivity list, or one or more **wait** statements to suspend the process. (It is not legal to have both a sensitivity list and a **wait** statement.)

Concurrent and Sequential VHDL

Understanding the fundamental difference between concurrent and sequential statements in VHDL is important to making effective use of the language. The diagrams of Figure 2-12 illustrate the basic difference between these two types of statements:

Figure 2-12: VHDL allows both concurrent and sequential statements to be entered.

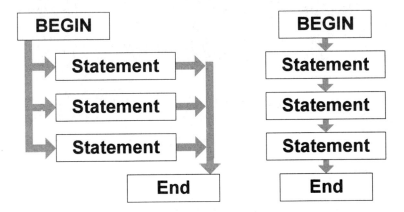

All statements in the concurrent area are executed at the same time, and there is no significance to the order in which concurrent statements are entered.

The left-most diagram illustrates how concurrent statements are executed in VHDL. Concurrent statements are those statements that appear between the **begin** and **end** statements of a VHDL architecture. This area of your VHDL architecture is what is known as the concurrent area. In VHDL, all statements in the concurrent area are executed at the same time, and there is no significance to the order in which the statements are entered.

The right-most diagram shows how sequential VHDL statements are executed. Sequential statements are executed one after the other in the order that they appear between the **begin** and **end** statements of a VHDL architecture.

The interaction of concurrent and sequential statements is illustrated in the example below. While the **if-elsif-end-if** statements in the body of the process are executed sequentially (i.e., one after the other), the body of the process is treated by VHDL as a single concurrent statement and is executed at the same time as all other concurrent statements in the simulation.

You can think of the process as one large concurrent statement that itself contains a series of sequential statements.

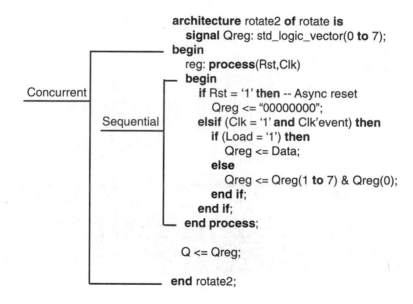

Note:
Writing a description of a circuit using the sequential programming features of VHDL (statements entered within processes and subprograms) does not necessarily mean that the circuit being described is sequential in its operation. Sequential circuits require some sort of internal memory (such as one or more flip-flops or latches) to operate, and a VHDL process or subprogram may or may not imply such memory elements. As we will see

in later chapters, it is actually quite common to describe strictly combinational circuits—circuits having no memory and, hence, no sequential behavior—using sequential statements within processes and subprograms.

Signals and Variables

There are two fundamental types of objects used to carry data from place to place in a VHDL design description: signals and variables. You can use variables to simplify sequential statements (within processes, procedures and functions), but you must always use signals to carry information between concurrent elements of your design (such as between two independent processes).

It is useful to think of signals as wires, as in a schematic, and variables as being more similar to variables found in software programming languages.

Signals and variables are described in more detail in a later chapter. For now, it is useful to think of signals as wires (as in a schematic) and variables as temporary storage areas similar to variables in a traditional software programming language.

In many cases, you can choose whether to use signals or variables to perform the same task. For your first design efforts you should probably stick to using signals, unless you fully understand the consequences of using variables for your particular application.

Using a Procedure to Describe Registers

As we have seen from the first version of the 8-bit shifter, describing registered logic using processes requires you to follow some established conventions (if you intend to synthesize the design) and to consider the behavior of the entire circuit. In the shifter design description previously shown, the registers were implied by the placement and use of statements such as if **Clk = '1' and Clk'event**. Assignment statements subject to that clause resulted in D-type flip-flops being implied for the signals.

For smaller circuits, this mixing of combinational logic functions and registers is fine and not difficult to understand. For larger circuits, however, the complexity of the system being described can make such descriptions hard to manage, and the results of synthesis can often be confusing. For these circuits, you might choose to retreat to more of a dataflow level of abstraction and to clearly define the boundaries between registered and combinational logic.

Many VHDL users find it easier to write concurrent, dataflow-level statements for registered logic rather than use processes.

One way to do this is to remove the process from your design and replace it with a series of concurrent statements representing the combinational and registered portions of the circuit. You can do this using either procedures or lower-level components to represent the registers. The following VHDL design description uses a register procedure to describe the same shifter circuit previously described:

```
architecture shift3 of shift is
    signal D,Qreg: std_logic_vector(0 to 7);
begin

    D <= Data when (Load = '1') else
            Qreg(1 to 7) & Qreg(0);

    dff(Rst, Clk, D, Qreg);

    Q <= Qreg;

end rotate3;
```

In the **shift3** version of the design description above, the behavior of the D-type flip-flop has been placed in an external procedure, **dff()**, and intermediate signals have been introduced to more clearly describe the separation between the combinational and registered parts of the circuit. Figure 2-13 helps to illustrate this separation.

In this example, the combinational logic of the counter has been written in the form of a single concurrent signal assignment, while the registered operation of the counter's output has been described using a call to a procedure named **dff**.

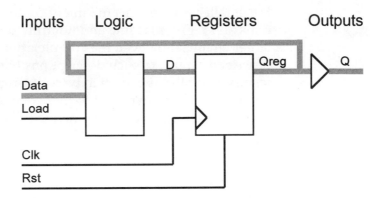

Figure 2-13: When using a dataflow level of abstraction, you clearly distinguish the registered and combinational parts of your design.

What does the **dff** procedure look like? The following is one possible procedure for a D-type flip-flop:

```
procedure dff (signal Rst, Clk: in std_logic;
               signal D: in std_logic_vector(0 to 7);
               signal Q: out std_logic_vector(0 to 7)) is
begin
  if Rst = '1' then
     Q <= "00000000";
  elsif Clk = '1' and Clk'event then
     Q <= D;
  end if;
end dff;
```

Warning: Bad Code!

The above procedure is actually rather poorly written, as it assumes a specific width (eight bits) for the D and Q parameters. A more flexible and reusable procedure could be written that makes use of attributes (such as 'length) to determine the actual number of bits of the signals within the procedure.

A procedure can look very much like a process...

Notice that this procedure has a striking resemblance to the process statement presented earlier. The same **if-then-elsif** structure used in the process is used to describe the behavior of the registers. Instead of a sensitivity list, however, the procedure has a parameter list describing the inputs and outputs of the procedure.

The parameters defined within a procedure or function definition are called its *formal parameters*. When the procedure or function is executed in simulation, the formal parameters are replaced by the values of the *actual parameters* specified when the procedure or function is used. If the actual parameters being passed into the procedure or function are signal objects, then the **signal** keyword can be used (as shown above) to ensure that all information about the signal object, including its value and all of its attributes, is passed into the procedure or function.

Using a Component to Describe Registers

Another common approach to dataflow-level design is to make use of components for the registered elements of the circuit.

It is important to note that if you wish to use procedures to describe your registers, you will need to make sure the synthesis tool you are using allows this. While most VHDL synthesis tools support the use of procedures for registers, some do not, while others have severe restrictions on the use of procedures. If you are not sure, or if you want to have the most portable design descriptions possible, you should replace procedures such as this with components, as in the following example:

```
architecture shift4 of shift is
    component dff
        port(Rst, Clk: std_logic;
              D: in std_logic_vector(0 to 7);
              Q: out std_logic_vector(0 to 7));
    end component;

    signal D, Qreg: std_logic_vector(0 to 7);

begin

    D <= Data when (Load = '1') else
            Qreg(1 to 7) & Qreg(0);
    REG1: dff port map(Rst, Clk, D, Qreg);
    Q <= Qreg;

end shift4;
```

An entity and architecture pair describing exactly the same behavior as the **dff** procedure is shown below:

```
library ieee;
use  ieee.std_logic_1164.all;

entity dff is
   port (Rst, Clk: in std_logic;
           D: in std_logic_vector(0 to 7);
           Q: out std_logic_vector(0 to 7));
end dff;

architecture behavior of dff is
begin
   process (Rst, Clk)
   begin
     if Rst = '1' then
       Q <= "00000000";
     elsif Clk = '1' and Clk'event then
       Q <= D;
     end if;
   end process;
end behavior;
```

Structural VHDL

Structure-level design methods can be useful for managing the complexity of a large design description.

The structure level of abstraction is used to combine multiple components to form a larger circuit or system. As such, structure can be used to help manage a large and complex design, and structure can make it possible to reuse components of a system in other design projects.

Because structure only defines the interconnections between components, it cannot be used to completely describe the function of a circuit; at the lowest level, all components of your system must be described using behavioral and/or dataflow levels of abstraction.

To demonstrate how the structure level of abstraction can be used to connect lower-level circuit elements into a larger circuit, we will connect the comparator and shift register circuits into a larger circuit as shown in Figure 2-14.

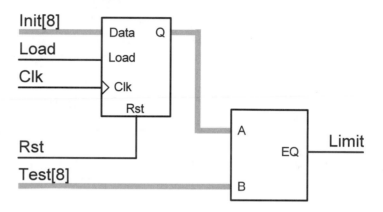

Figure 2-14: Our sample design has now grown to include both the comparator and shifter.

Notice that we have drawn this diagram in much the same way you might enter it into a schematic capture system. Structural VHDL has many similarities with schematic-based design, as we will see.

Design Hierarchy

Structural VHDL descriptions are quite similar in format to schematic netlists.

When you write structural VHDL, you are in essence writing a textual description of a schematic *netlist* (a description of how the components on the schematic are connected by wires, or *nets*). In the world of schematic entry tools, such netlists are usually created for you automatically by the schematic editor. When writing VHDL, you enter the same sort of information by hand. (Note: many schematic capture tools in existence today are capable of writing netlist information in the form of a VHDL source file. This can save you a lot of time if you are used to drawing block diagrams in a schematic editor.)

When you use components and wires (signals, in VHDL) to connect multiple circuit elements together, it is useful to think of your new, larger circuit in terms of a *hierarchy* of components. In this view, the top-level drawing (or top-level VHDL entity and architecture) can be seen as the highest level in a hierarchy tree, as shown in Figure 2-15.

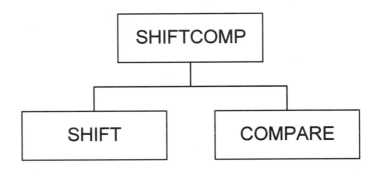

Figure 2-15: The hierarchy of a design can be represented as a tree of connected components.

Larger circuits can be constructed from smaller building blocks...

In this example, we have introduced a new top-level component (called **shiftcomp**) that references the two lower-level components **shift** and **compare**. Because the new **shiftcomp** design entity can itself be viewed as a component, and considering the fact that any component can be referenced more than once, we quickly see how very large circuits can be constructed from smaller building blocks.

The following VHDL source file describes our complete circuit using structural VHDL statements (component declarations and component instantiations) to connect together the **compare** and **shift** portions of the circuit:

```
library ieee;
use ieee.std_logic_1164.all;
entity shiftcomp is port(Clk, Rst, Load: in std_ulogic;
                  Init: in std_ulogic_vector(0 to 7);
                  Test: in std_ulogic_vector(0 to 7);
                  Limit: out std_ulogic);
end shiftcomp;

architecture structure of shiftcomp is
```

```
component compare
    port(A, B: in std_ulogic_vector(0 to 7); EQ: out std_ulogic);
end component;


component shift
    port(Clk, Rst, Load: in std_ulogic;
        Data: in std_ulogic_vector(0 to 7);
        Q: out std_ulogic_vector(0 to 7));
end component;

signal Q: std_ulogic_vector(0 to 7);

begin

    COMP1: compare port map (A=>Q, B=>Test, EQ=>Limit);
    SHIFT1: shift port map (Clk=>Clk, Rst=>Rst, Load=>Load, Data=>Init,
                            Q=>Q);

end structure;
```

Note:

In the above context, the VHDL symbol => is used to associate the signals within an architecture to ports defined within the lower-level component. See Chapter 8, Partitioning Your Design.

VHDL provides many ways to describe component interconnections.

There are many ways to express the interconnection of components and to improve the portability and reusability of those components. We will examine these more advanced uses of components in a later chapter.

Test Benches

At this point, our sample circuit is complete and ready to be processed by a synthesis tool. Before processing the design, however, we should take the time to verify that it actually does what it is intended to do, by running a simulation.

Simulating a circuit such as this one requires that we provide more than just the design description itself. To verify the proper operation of the circuit over time in response to input stimulus, we will need to write a test bench.

You can think of a test bench as a virtual circuit tester...

The easiest way to understand the concept of a test bench is to think of it as a virtual circuit tester. This tester, which you will describe in VHDL, applies stimulus to your design description and (optionally) verifies that the simulated circuit does what it is intended to do.

Figure 2-16 graphically illustrates the relationship between the test bench and your design description, which is called the *unit under test*, or *UUT*.

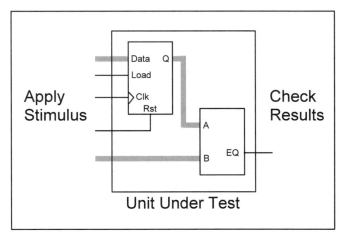

Figure 2-16: A test bench is a top-level VHDL description of a sequence of tests to be applied to your design.

To apply stimulus to your design, your test bench will probably be written using one or more sequential processes, and it will use a series of signal assignments and **wait** statements to describe the actual stimulus. You will probably use VHDL's looping features to simplify the description of repetitive stimulus (such as the system clock), and you may also use VHDL's file and record features to apply stimulus in the form of test vectors.

To check the results of simulation, you will probably make use of VHDL's **assert** feature, and you may also use the text I/O features to write the simulation results to a disk file for later analysis.

Developing a compre-hensive test bench can be a large-scale project in itself...

For complex design descriptions, developing a comprehensive test bench can be a large-scale project in itself. In fact, it is not unusual for the test bench to be larger and more complex than the underlying design description. For this reason, you should plan your project so that you have the time required to develop the test bench in addition to developing the circuit being tested. You should also plan to create test benches that are re-usable, perhaps by developing a master test bench that reads test data from a file.

When you create a test bench for your design, you use the structural level of abstraction to connect your lower-level (previously top-level) design description to the other parts of the test bench, as shown in Figure 2-17.

Figure 2-17: During simulation, the test bench becomes the top level of your design.

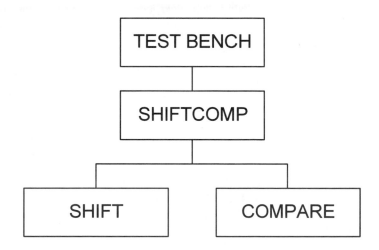

63

Sample Test Bench

The following VHDL source statements, with explanatory comments, describe a simple test bench for our sample circuit:

Even a simple test bench can require more lines of VHDL statements than the underlying design description...

```vhdl
library ieee;
use ieee.std_logic_1164.all;

entity testbnch is                    -- No ports needed in a
end testbnch;                         -- testbench

use work.shiftcomp;
architecture behavior of testbnch is
    component shiftcomp is            -- Declares the lower-level
        port(Clk, Rst, Load: in std_logic;   -- component and its ports
            Init: in std_logic_vector(0 to 7);
            Test: in std_logic_vector(0 to 7);
            Limit: out std_logic);
    end component;
    signal Clk, Rst, Load: std_logic;         -- Introduces top-level signals
    signal Init: std_logic_vector(0 to 7);    -- to use when testing the
    signal Test: std_logic_vector(0 to 7);    -- lower-level circuit
    signal Limit: std_logic;
begin
    DUT: shiftcomp port map                   -- Creates an instance of the
        (Clk, Rst, Load, Init, Test, Limit);-- lower-level circuit (the
                                              -- unit under test)

    clock: process
        variable clktmp: std_logic := '0';    -- This process sets up a
    begin                                     -- background clock of 100 ns
        clktmp := not clktmp;                 -- period.
        Clk <= clktmp;
        wait for 50 ns;
    end process;

    stimulus: process                         -- This process applies
    begin                                     -- stimulus to the design
        Rst <= '0';                           -- inputs, then waits for some
        Load <= '1';                          -- amount of time so we can
        Init <= "00001111";                   -- observe the results during
        Test <= "11110000";                   -- simulation.
        wait for 100 ns;
        Load <= '0';
        wait for 600 ns;
    end process;
end behavior;
```

More advanced applications of test benches can be found in Chapter 9: *Writing Test Benches*.

What We've Learned So Far

Wow! We've covered a lot of ground. So far, we've learned about VHDL design elements, different levels of abstraction, and concurrent and sequential VHDL. In addition, we've seen examples of hierarchy, a test bench, and we've talked about synthesizable VHDL. If you have questions, don't worry. All of these topics will be covered in more detail in later chapters.

For now, pour yourself a cup of your favorite hot drink and let's continue our adventure together by...

3. Exploring Objects and Data Types

VHDL includes a number of language elements that can be used to represent and store data...

VHDL includes a number of language elements, collectively called *objects*, that can be used to represent and store data in the system being described. The three basic types of objects that you will use when entering a design description for synthesis or creating functional tests (in the form of a test bench) are *signals*, *variables* and *constants*. Each object that you declare has a specific *data type* (such as **bit** or **integer**) and a unique set of possible values.

The values that an object can take will depend on the definition of the type used for that object. For example, an object of type **bit** has only two possible values, **'0'** and **'1'**, while an object of type **real** has many possible values (floating point numbers within a precision and range defined by the VHDL standard and by the specific simulator you are using).

When an explicit value is specified (such as when you are assigning a value to a signal or variable, or when you are passing a value as a parameter to a subprogram), that value is represented in the form of a *literal*.

Signals

Signals are similar to wires on a schematic, and can be used to interconnect concurrent elements of the design.

Signals are objects that are used to connect concurrent elements (such as components, processes and concurrent assignments), similar to the way that wires are used to connect components on a circuit board or in a schematic. Signals can be declared globally in an external package or locally within an architecture, block or other declarative region.

To declare a signal, you write a **signal** statement such as the following:

```
architecture arch1 of my_design is
    signal Q: std_logic;
begin
    . . .
end arch1;
```

In this simple example, the signal **Q** is declared within the declaration section of the **arch1** architecture. At a minimum, a signal declaration must include the name of the signal (in this case **Q**) and its type (in this case the standard type **std_logic**). If more than one signal of the same type is required, multiple signal names can be specified in a single declaration:

```
architecture arch2 of my_design is
    signal Bus1, Bus2: std_logic_vector(7 downto 0);
begin
    . . .
end declare;
```

In the first example above, the declaration of **Q** was entered in the declaration area of architecture **arch1**. Thus, the signal **Q** will be visible anywhere within the **arch1** design unit, but it will not be visible within other design units. To make the signal **Q** visible to the entire design (a *global signal*), you would have to move the declaration into an external package, as shown below:

```
package my_package is
    signal Q: std_logic;    -- Global signal
end my_package;
    . . .
```

```
use work.my_package.Q;   -- Make Q visible to the architecture
architecture arch1 of my_design is
begin
  . . .
end arch1;
```

In this example, the declaration for **Q** has been moved to an external package, and a **use** statement has been specified, making the contents of that package visible to the subsequent architecture. (For more information about packages and **use** statements, refer to Chapter 8, *Partitioning Your Design*).

Signal Initialization

Signals can be initialized to a specific value.

In addition to creating one or more signals and assigning them a type, the signal declaration can also be used to assign an initial value to the signal, as shown below:

```
signal BusA: std_logic_vector(15 downto 0) := (others => 'Z');
```

This particular initialization uses a special kind of assignment, called an *aggregate assignment*, to assign all signals of the array **BusA** to an initial value of 'Z'. (The 'Z' value is defined in the IEEE 1164 standard as a high-impedance state.)

Initialization values are useful for simulation modeling, but they are not recommended for design descriptions that will be processed by synthesis tools. Synthesis tools typically ignore initialization values because they cannot assume that the target hardware will power up in a known state.

Using Signals

You will use signals in VHDL in two primary ways. First, if you want signals to carry information between different functional parts of your design, such as between two components, you will probably use them in a way similar to the following:

```
library ieee;
use ieee.std_logic_1164.all;
entity shiftcomp is port(Clk, Rst, Load: in std_logic;
                         Init: in std_logic_vector(0 to 7);
```

```
                                    Test: in std_logic_vector(0 to 7);
                                    Limit: out std_logic);
        end shiftcomp;

        architecture structure of shiftcomp is

            component compare
                port(A, B: in std_logic_vector(0 to 7); EQ: out bit);
            end component;

            component shift
                port(Clk, Rst, Load: in std_logic;
                    Data: in std_logic_vector(0 to 7);
                    Q: out std_logic_vector(0 to 7));
            end component;

            signal Q: std_logic_vector(0 to 7);

        begin

            COMP1: compare port map (Q, Test, Limit);
            SHIFT1: shift port map (Clk, Rst, Load, Init, Q);

        end structure;
```

This example (described in Chapter 2, *A First Look At VHDL*) declares the signal **Q** within the architecture, then uses **Q** to connect the two components **COMP1** and **SHIFT1** together.

Signals can also be used to describe complex logical expressions.

A second way of using signals is demonstrated by the following example in which signals are used within logic expressions and are assigned values directly (in this case within a process):

```
library ieee;
use ieee.std_logic_1164.all;
entity synch is
    port (Rst, Clk, Grant, nSelect: in std_logic;
        Request: out std_logic);
end synch;

architecture dataflow of synch is
    signal Q1, Q2, Q3, D3: std_logic;
begin
```

```
dff: process (Rst, Clk)
begin
  if Rst = '1' then
      Q1 <= '0';
      Q2 <= '0';
      Q3 <= '0';
  elsif Clk = '1' and Clk'event then
      Q1 <= Grant;
      Q2 <= Select;
      Q3 <= D3;
  end if;
end process;

D3 <= Q1 and Q3 or Q2;
Request <= Q3;

end dataflow;
```

This example (which is a simplified synchronizer circuit) uses three signals, **Q1**, **Q2** and **Q3**, to represent register elements, with the signal **D3** being used as an intermediate signal representing a combinational logic function connecting the outputs of registers **Q1**, **Q2** and **Q3** to the input of **Q3**. The final assignment assigns the **Q3** register output to the **Request** output port. The register behavior is encapsulated into a process, **dff**, simplifying the concurrent statements that follow.

It is important to note that there is no significance to the order in which these concurrent statements occur. Like wires drawn between symbols on a schematic, signals assigned and used within a VHDL architecture are independent of each other and are not position dependent.

Variables

Variables can be used only within processes, functions and procedures.

Variables are objects used to store intermediate values between sequential VHDL statements. Variables are only allowed in processes, procedures and functions, and they are always local to those functions.

Note:

The 1076-1993 language standard adds a new type of global variable that has visibility between different processes and subprograms. Global variables are not generally supported in synthesis tools and are not discussed in this book.

Variables in VHDL are much like variables in a conventional software programming language. They immediately take on and store the value assigned to them (this is not true of signals, as described in Chapter 6, *Understanding Sequential Statements*), and they can be used to simplify a complex calculation or sequence of logical operations.

The following example is a modified version of the synchronizer circuit presented in the previous section:

```
library ieee;
use ieee.std_logic_1164.all;
entity synch is
    port (Rst, Clk, Grant, nSelect: std_ulogic;
          Request: std_ulogic);
end synch;
```

Variables can represent intermediate storage elements which may or may not imply registers.

```
architecture behavior of synch is
begin
    process(Rst, Clk)
        variable Q1, Q2, Q3: std_ulogic;
    begin
        if Rst = '1' then   -- Async reset
            Q1 := '0'; Q2 := '0'; Q3 := '0';
        elsif (Clk = '1' and Clk'event) then
            Q1 := Grant;
            Q2 := Select;
            Q3 := Q1 and Q3 or Q2;
        end if;
        Request <= Q3;
    end process;
end behavior;
```

In this version of the synchronizer, a single process is again used to describe the behavior of the three commonly-clocked register elements. But in this case, the connections between the three registers are represented by variables that are local to the process, and the result (the output of register **Q3**) is then assigned to the output port **Request**. This version of the design will probably not work as intended, because the registered behavior of **Q1** and **Q2** will be "short circuited" by the fact that variables were used.

Because variables do not always result in registers being generated within otherwise clocked processes, you must be very careful when using them. The important distinctions between signals and variables are covered in more detail in a later chapter.

Constants

Constants are useful for representing commonly-used values of specific types.

Constants are objects that are assigned a value once, when declared, and do not change their value during simulation. Constants are useful for creating more readable design descriptions, and they make it easier to change the design at a later time. The following code fragment provides a few examples of constant declarations:

```
architecture sample1 of consts is
    constant SRAM: bit_vector(15 downto 0) := X"F0F0";
    constant PORT: string  := "This is a string";
    constant error_flag: boolean := True;
begin
    . . .
    process( . . .)
       constant CountLimit: integer := 205;
    begin
      . . .
    end process;

end arch1;
```

Constant declarations can be located in any declaration area in your design description. If you want to create constants that are global to your design description, then you will place the constant declarations into external packages (see Chapter 8, *Partitioning Your Design*). If a constant will be used only within one segment of your design, you can place the constant declaration within the architecture, block, process or subprogram that requires it.

Literals

Literals do not always have an explicit data type...

Explicit data values that are assigned to objects or used within expressions are called *literals*. Literals represent specific values, but they do not always have an explicit type. (For example, the literal **'1'** could represent either a **bit** data type or a **character**.) Literals do, however, fall into a few general categories.

Character Literals

Character literals are 1-character ASCII values that are enclosed in single-quotes, such as the values **'1'**, **'Z'**, **'$'** and **':'**. The data type of the object being assigned one of these values (or the type implied by the expression in which the value is being used) will dictate whether a given character literal is valid. The literal value **'$'**, for example, is a valid literal when assigned to a **character** type object, but it is not valid when assigned to a **std_logic** or **bit** data type.

String Literals

String literals are collections of one or more ASCII characters enclosed in double-quote characters. String literals may contain any combination of ASCII characters, and they may be assigned to appropriately sized arrays of single-character data types (such as **bit_vector** or **std_logic_vector**) or to objects of the built-in type **string**.

Bit String Literals

Bit string literals are special forms of string literals that are used to represent binary, octal, or hexadecimal numeric data values.

When representing a binary number, a bit string literal must be preceded by the special character 'B', and it may contain only the characters '0' and '1'. For example, to represent a decimal value of 36 using a binary format bit string literal, you would write **B"100100"**.

Bit string literals can be represented using non-binary formats...

When representing an octal number, the bit string literal must include only the characters '0' through '7', and it must be preceded by the special character 'O', as in **O"446"**.

When representing a hexadecimal value, the bit string literal must be preceded by the special character 'X', and it may include only the characters '0' through '9' and the characters 'A' through 'F', as in **X"B295"**. (Lower-case characters are also allowed, so 'a' through 'f' are also valid.)

The underscore character **'_'** may also be used in bit string literals as needed to improve readability. The following are some examples of bit string literals representing a variety of numeric values:

B"0111_1101"   (decimal value 253)

O"654"   (decimal value 428)

O"146_231"   (decimal value 52,377)

X"C300"   (decimal value 49,920)

Note:
*In VHDL standard 1076-1987, bit string literals are only valid for the built-in type **bit_vector**. In 1076-193, bit string literals can be applied to any string type, including **std_logic_vector**.*

Numeric Literals

There are two basic forms of numeric literals in VHDL, *integer literals* and *real literals*.

Integer literals are entered as you would expect, as decimal numbers preceded by an optional negation character ('-'). The range of integers supported is dependent on your particular simulator or synthesis tool, but the VHDL standard does specify a minimum range of -2,147,483,647 to +2,147,483,647 (32 bits of precision, including the sign bit).

Real literals are entered using an extended form that requires a decimal point. For large numbers, scientific notation is also allowed using the character 'E', where the number to the left of the 'E' represents the mantissa of the real number, while the number to the right of the 'E' represents the exponent. The following are some examples of real literals:

5.0

-12.9

1.6E10

1.2E-20

The actual range of a real number literal is defined as a minimum range in the VHDL specification.

The minimum and maximum values of real numbers are defined by the simulation tool vendor, but they must be at least in the range of -1.0E38 to +1.0E38 (as defined by the standard). Numeric literals may not include commas, but they may include underscore characters ("_") to improve readability, as in:

1_276_801   -- integer value 1,276,801

Type checking is strict in VHDL, and this includes the use of numeric literals. It is not possible, for example, to assign an integer literal of **9** to an object of type **real**. (You must instead enter the value as **9.0**.)

Based Literals

Based literals are another form of integer or real values, but they are written in non-decimal form. To specify a based literal, you precede the literal with a base specification (such as 2, 8, or 16) and enclose the non-decimal value with a pair of '#' characters as shown in the examples below:

2#10010001#   (integer value 145)

16#FFCC#   (integer value 65,484)

2#101.0#E10   (real value 5,120.0)

Physical Literals

Physical literals are special types of literals used to represent physical quantities such as time, voltage, current, distance, etc. Physical literals include both a numeric part (expressed as an integer) and a unit specification. Physical types will be described in more detail later in this chapter. The following examples show how physical literals can be expressed:

300 ns   (300 nanoseconds)

900 ps   (900 picoseconds)

40 ma   (40 milliamps)

Types and Subtypes

The VHDL 1076 specification describes four classes of data types:

VHDL's data types fall into four general catagories...

- **Scalar types** represent a single numeric value or, in the case of enumerated types, an enumeration value. The standard types that fall into this class are integer, real (floating point), physical, and enumerated types. All of these basic types can be thought of as numeric values.

- **Composite types** represent a collection of values. There are two classes of composite types: arrays containing elements of the same type, and records containing elements of different types.

- **Access types** provide references to objects in much the same way that pointer types are used to reference data in software programming languages.

- **File types** reference objects (typically disk files) that contain a sequence of values.

Each data type has a uniquely-defined set of possible values.

Each type in VHDL has a defined set of values. For example, the value of an integer data type has a defined range of at least -2147483647 to +2147483647. In most cases you will only be interested in a subset of the possible values for a type, so VHDL provides the ability to specify a constraint whenever an object of a given type is declared. The following declaration creates an object of type integer that is constrained to the positive values of 0 to 255:

signal ShortInt: integer range 0 to 255;

Subtypes can be used to constrain an existing type to a subset of its values.

VHDL also provides a feature called a **subtype**, allowing you to declare an alternate data type that is a constrained version of an existing type. For example, the declaration

subtype SHORT integer range 0 to 255;

creates an alternate scalar type with a constrained range. Because **SHORT** is a subtype of integer, it carries with it all operations available for the integer base type.

The four classes of data types are discussed in more detail below.

Scalar Types

Scalar types are those types that represent a single value, and are ordered in some way so that relational operations (such as greater than, less than, etc.) can be applied to them. These types include the obvious numeric types (integer and real) as well as less obvious enumerated types such as Boolean and Character.

The table of Figure 3-1 lists the built-in scalar types defined in VHDL Standard 1076.

Data Type	Typical Values	Comment
Bit	'1', '0'	Defined as an enumerated type.
Boolean	True, False	The result of any comparison operation. Also an enumerated type.
Integer	0, 1, 2, -32,129, etc.	Minimum range of -2147483647 to +2147483647, inclusive.
Character	'a', 'b', '0', '1', '2', '$', '@', etc.	ISO 8859-1 character set.
Real	Floating point number	Minimum range of -1.0E38 to_+1.0E38.
Severity_level	NOTE, ERROR	Enumerated type used in severity clause of report.
Time	100 ns	A physical type.

Figure 3-1. *Scalar types have an implied order that allows relational operators to be applied to them.*

Bit Type

The **bit** data type is the most fundamental representation of a wire in VHDL. The **bit** type has only two possible values, '0' and '1', that can be used to represent logical 0 and 1 values (respectively) in a digital system. The following example uses **bit** data types to describe the operation of a full adder:

```
entity fulladder is
    port (X: in bit;
        Y: in bit;
        Cin: in bit;
        Cout: out bit;
        Sum: out bit);
end fulladder;

architecture concurrent of fulladder is
begin

    Sum <= X xor Y xor Cin;
```

Cout <= (X **and** Y) **or** (X **and** Cin) **or** (Y **and** Cin);

end concurrent;

The bit data type supports the following operations: and, or, nand, nor, xor, xnor, not, =, /=, <, <=, >, and >=.

Note:
*The IEEE 1164 specification describes an alternative to **bit** called **std_ulogic**. Std_ulogic has nine possible values, allowing the values and states of wires (such as high-impedence, unknown, etc.) to be more accurately described. (See Chapter 4, Using Standard Logic.)*

Boolean Type

The **Boolean** type has two possible values, **True** and **False**. Like the **bit** data type, the **Boolean** type is defined as an enumerated type. The **Boolean** type does not have any implied width; it is simply the result of a logical test (such as a comparison operation or an **if** statement) or the expression of some logical state (such as in the assignment, *ErrorFlag <= True;*).

Integer Type

An integer type includes integer values in a specified range. The only predefined integer type is **integer**. Integer types have a minimum default range of -2147483647 to +2147483647, inclusive. However, you can restrict that value with a range constraint and/or declare a new integer subtype with a range constrained range, as in the following example:

subtype byteint integer **range** 0 **to** 255;

The predefined subtype **natural** restricts integers to the range of 0 to the specified (or default) upper range limit. The predefined subtype **positive** restricts integers to the range of 1 to the specified (or default) upper range limit.

An alternative to VHDL's built-in integer type is provided in Standard 1076.3.

An alternative to the integer data type is provided with IEEE Standard 1076.3. This standard defines the standard data types **signed** and **unsigned**, which are array types (based on the

IEEE 1164 9-valued data types) that have properties of both array (composite) and numeric (scalar) data types. Like an array, you can perform shifting and masking operations on them and, like integers, you can perform arithmetic operations on them. More information about the IEEE 1076.3 data types can be found in Chapter 4, Using Standard Logic.

Real (Floating Point) Types

Floating point types are used to approximate real number values. The predefined floating point type provided in VHDL is called **real**. It has possible values in the range of at least -1.0E38 to +1.0E38.

The following declaration decribes a signal of type **real** that has been initialized to a real value of 4589.3:

```
signal F0: real := 4589.3;
```

The real data type supports the following operations: =, /=, <, <=, >, >=, +, -, abs, +, -, *, and /.

Note:

Floating point types have little use in synthesizable designs, as no synthesis tool available today will accept them.

Character Type

VHDL's character data type is similar to the character types you might be familiar with from software languages. Characters can be used to represent string data (such as you might use in a test bench), to display messages during simulation, or to represent actual data values in your design description. Unlike many software languages, character values in VHDL have no explicit value. This means that they cannot be simply mapped onto numeric data types or assigned directly to arrays of bits.

The character values defined in the 1076-1987 standard package are defined in Figure 3-2.

Figure 3-3. The *character literals supported in VHDL are defined as enumerated types, and therefore have no explicit numeric value.*

NUL,	SOH,	STX,	ETX,	EOT,	ENQ,	ACK,	BEL,
BS,	HT,	LF,	VT,	FF,	CR,	SO,	SI,
DLE,	DC1,	DC2,	DC3,	DC4,	NAK,	SYN,	ETB,
CAN,	EM,	SUB,	ESC,	FPS,	GSP,	RSP,	USP,
' ',	'!',	'"',	'#',	'$',	'%',	'&',	''',
'(',	')',	'*',	'+',	',',	'-',	'.',	'/',
'0',	'1',	'2',	'3',	'4',	'5',	'6',	'7',
'8',	'9',	':',	';',	'<',	'=',	'>',	'?',
'@',	'A',	'B',	'C',	'D',	'E',	'F',	'G',
'H',	'I',	'J',	'K',	'L',	'M',	'N',	'O',
'P',	'Q',	'R',	'S',	'T',	'U',	'V',	'W',
'X',	'Y',	'Z',	'[',	'\',	']',	'^',	'_',
'`',	'a',	'b',	'c',	'd',	'e',	'f',	'g',
'h',	'i',	'j',	'k',	'l',	'm',	'n',	'o',
'p',	'q',	'r',	's',	't',	'u',	'v',	'w',
'x',	'y',	'z',	'{',	'|',	'}',	'~',	DEL

VHDL Standard 1076-1993 extends the character data type to the 256-character ISO 8859 character set.

The IEEE 1076-1993 specification extends this character set to the 256-character ISO 8859 standard.

Once again, there is no specific numeric value associated with a given character literal in VHDL. (You cannot, for example, assign a character literal to an 8-bit array without providing a type conversion function that assigns unique array values—such as ASCII values—to the target array for each character value.)

The character data type is an enumerated type. However, there is an implied ordering (as listed above).

Severity_Level Type

*Type **severity_level** is only used in VHDL's **assert** statement.*

Severity_level is a special data type used in the **report** section of an **assert** statement. There are four possible values for an object of type **severity_level**: **note**, **warning**, **error** and **failure**. You might use these severity levels in your test bench, for example, to instruct your simulator to stop processing when an error (such as a test vector failure) is encountered during a simulation run. The following **assert** statement makes use of the FAILURE severity level to indicate that the simulator should halt processing if the specified condition evaluates false:

```
assert (error_flag = '1')
    report "There was an error; simulation has halted."
    severity FAILURE;
```

Time and Other Physical Types

*Physical types such as **time** define specific units that are are multiples of a base unit. Subtypes can be used to constrain an existing type to a subset of its values.*

Time is a standard data type that falls into the catagory of physical types in VHDL. Physical types are those types that are used for measurement. They are distinquished by the fact that they have units of measure, such as (in the case of time) seconds, nanoseconds, etc. Each unit in the physical type (with the exception of the base unit) is based on some multiple of the preceding unit. The definition for type time, for example, might have been written as follows (the actual definition is implementation-dependent):

```
type time is range -2147483647 to 2147483647
    units
        fs;
        ps  = 1000 fs;
        ns  = 1000 ps;
        us  = 1000 ns;
        ms  = 1000 us;
        sec = 1000 ms;
        min = 60 sec;
        hr  = 60 min;
    end units;
```

Enumerated Types

As we have seen, enumerated types are used to describe (internally) many of the standard VHDL data types. You can also use enumerated types to describe your own unique data types. For example, if you are describing a state machine, you might want to make use of an enumerated type to represent the various states of the machine, as in the following example:

Enumerated types can be used to describe high-level design concepts using symbolic values.

```
architecture FSM of VCONTROL is
    type states is (StateLive,StateWait,StateSample,StateDisplay);
    signal current_state, next_state: states;
begin
    . . .
    -- State transitions:
    STTRANS: process(current_state,Mode,VS,ENDFR)
    begin
        case current_state is
            when StateLive =>    -- Display live video on the output
            . . .
            when StateWait =>    -- Wait for vertical sync
            . . .
            when StateSample => -- Sample one frame of video
            . . .
            when StateDisplay => -- Display the stored frame
            . . .
        end case;
    end process;
end FSM;
```

In this example (the control logic for a video frame grabber, described in detail in Chapter 6, *Understanding Sequential Statements*), an enumerated type (**states**) is defined in the architecture, and two signals (**current_state** and **next_state**) are declared for use in the subsequent state machine description. Using enumerated types in this way has two primary advantages: first, it is very easy to debug a design that uses enumerated types, because you can observe the symbolic type names during simulation; second, and perhaps more importantly for this state machine description, you can defer the actual encoding of the symbolic values until the time that you implement the design in hardware.

Synthesis tools generally recognize the use of enumerated types in this way and can perform special optimizations, assigning actual binary values to each symbolic name during synthesis. Synthesis tools generally also allow you to override the encoding of enumerated data types, so you have control over the encoding process.

Composite Types

Composite types include arrays (such as bit-vector and string) and records.

Data Type	Values	Comment
bit_vector	"00100101", "10", etc.	Array of **bit**
string	"Simulation failed!", etc.	Array of characters
records	Any collection of values	User defined composite data type.

Composite types are collections of one or more types of values. An array is a composite data type that contains items of the same type, either in a single dimension (such as a list of numbers or characters) or in multiple dimensions (such as a table of values). Records, on the other hand, define collections of possibly unrelated data types. Records are useful when you need to represent complex data values that require multiple fields.

Array Types

An array is a collection of one or more values or objects of the same type. Arrays are indexed by a number that falls into the declared range of the array.

The following is an example of an array type declaration:

type MyArray **is array** (15 **downto** 0) **of** std_ulogic;

*Arrays can be indexed using **to** or **downto** index ranges.*

This array type declaration specifies that the new type *MyArray* contains 16 elements, numbered downward from 15 to 0. Arrays can be given ranges that decrement from left to right (as shown) or increment (using the **to** keyword instead of **downto**). Index ranges do not have to begin or end at zero.

The index range (in this case *15 **downto** 0*) is what is known as the *index constraint*. It specifies the legal bounds of the array. Any attempt to assign values to, or read values from, an element outside the range of the array will result in an error during analysis or execution of the VHDL design description.

The index constraint for an array can specify an unbounded array using the following array range syntax:

```
type UnboundedArray is array (natural range <>) of std_ulogic;
```

Arrays can be declared with no specific range, creating an unbounded array.

This array type declaration specifies that the array UnboundedArray will have a index constraint matching the range of integer subtype *natural*, which is defined as 0 to the highest possible integer value (at least 2,147,483,647).

An array type is uniquely identified by the types (and constraints) of its elements, the number of elements (its range), and the direction and order of its indices.

VHDL also supports the use of multidimensional arrays.

Arrays can have multiple indices, as in the following example:

```
type multi is array(7 downto 0, 255 downto 0) of bit;
```

Note:
Multidimensional arrays are not generally supported in synthesis tools. They can, however, be useful for describing test stimulus, memory elements, or other data that require a tabular form.

The following example (a parity generator) demonstrates how array elements can be accessed, in this case within a loop:

```
entity parity10 is
   port(D: in array(0 to 9) of bit;
        ODD: out bit);
   constant WIDTH: integer := 10;
end parity10;

architecture behavior of parity10 is
begin
   process(D)
      variable otmp: Boolean;
   begin
```

```
        otmp := false;
        for i in 0 to D'length - 1 loop
          if D(i) = '1' then
            otmp := not otmp;
          end if;
        end loop;
        if otmp then
          ODD <= '1';
        else
          ODD <= '0';
        end if;
      end process;
end behavior;
```

It is important to consider the direction of an array before assigning values...

The direction of an array range has an impact on the index values for each element. For example, the following declarations:

```
    signal A: bit_vector(0 to 3);
    signal B: bit_vector(3 downto 0);
```

create two objects, **A** and **B**, that have the same width but different directions. The aggregate assignments:

```
    A <= ('1','0','1','0');
    B <= ('0','1','0','1');
```

are exactly identical to the assignments:

```
A(0) <= '1';
A(1) <= '0';
A(2) <= '1';
A(3) <= '0';

B(3) <= '0';
B(2) <= '1';
B(1) <= '0';
B(0) <= '1';
```

In this case, the arrays have the same contents when viewed in terms of their array indices. Assigning the value of **B** to **A**, as in:

```
    A <= B;
```

which would be exactly equivalent to the assignments:

```
A(0) <= B(3);
A(1) <= B(2);
A(2) <= B(1);
A(3) <= B(0);
```

The *leftmost* element of array **A** has an index of 0, while the leftmost value of array **B** has an index value of 1.

Record Types

A record is a composite type that has a value corresponding to the composite value of its elements. The elements of a record may be of unrelated types. They may even be other composite types, including other records. You can access data in a record either by referring to the entire record (as when copying the contents of one record object to another record object), or individually by referring to a *field name*. The following example demonstrates how you might declare a record data type consisting of four elements:

Record data types can be used to describe a collection of unrelated types.

```
type data_in_type is
    record
        ClkEnable: std_logic;
        Din: std_logic_vector(15 downto 0);
        Addr: integer range 0 to 255;
        CS: std_logic;
    end record;
```

The four names, **ClkEnable**, **Din**, **Addr** and **CS** are all field names of the record, representing data of specific types that can be stored as a part of the record. For example, an object of type **data_in_type** could be created and initialized with the following signal declaration:

```
signal test_record: data_in_type := ('0', "1001011011110011", 165, '1');
```

This initialization would be identical to the assignments:

```
test_record.ClkEnable <= '0';
 test_record.Din <= "1001011011110011";
test_record.Addr <= 165;
test_record.CS <= '1';
```

Records types are not generally synthesizable; however, they can be very useful when describing test stimulus. Examples shown in Chapter 9, *Writing Test Benches*, show how records can be used in combination with arrays to organize test stimulus.

Access and Incomplete Types

Access types are similar to pointers in a high-level software programming language.

Access types and incomplete types are used to create *data indirection* in VHDL. You can think of access types as being analogous to pointers in software programming languages such as C or Pascal. Incomplete types are required to create recursive types such as linked lists, trees and stacks. Access and incomplete types can be useful for creating dynamic representations of data (such as stacks), but they are not supported in today's synthesis tools. Refer to the IEEE VHDL Language Reference Manual for more information about these language features.

File Types

File types are very useful for writing test benches.

File types are very useful for writing test benches. File types differ in the VHDL 1076-1987 and 1076-1993 specifications. Discussions and examples of each are presented below.

VHDL 1076-1987 File Types

A file type is a special type of variable that contains sequential data. In the 1987 VHDL standard language, files are implicitly opened when they are declared, and it is not possible to explicitly close them. Objects of type file can be read from and written to using functions and procedures (**read**, **write**, and **endfile**) that are provided in the standard library. Additional functions and procedures for formating of data read from files is provided in the Text I/O library, which is also part of the 1076 standard. The built-in functions available for reading and writing files in VHDL (the 1987 specification) are:

Read(f, object)—Given a declared file and an object, read one field of data from the file into that object. When the read procedure is invoked, data is read from the file and the file is advanced to the start of the next data field in the file.

Write(f, object)—Given a declared file and an object, write the data contained in the object to the file.

Endfile(f)—Given a declared file, return a boolean **true** value if the current file marker is at the end of the file.

VHDL does not support random-access files.

Files in VHDL are sequential; there is no provision for opening a file and reading from a random location in that file, or for writing specific locations in a file.

To use an object of type **file**, you must first declare the type of its contents, as shown below:

```
type file_of_characters is file of character;
```

This declaration creates a new type, called **file_of_characters**, that consists of a sequence of character values. To use this file type, you would then create an object of type **file_of_characters**, as shown below:

```
file testfile: file_of_characters is in "TESTFILE.ASC";
```

This statement creates the object **testfile** and opens the indicated disk file. You can now use the built-in **read** procedure to access data in the file. A complete architecture that loops through a file and reads each character is shown below:

```
architecture sample87 of readfile is
begin
   Read_input: process
      type character_file is file of character;
      file cfile: character_file is in "TESTFILE.ASC";
      variable C: character;
      variable char_cnt: integer := 0;
   begin

      while not endfile(cfile) loop
         read (cfile, C) ;                -- Get a character from cfile into C
```

```
        char_cnt = char_cnt + 1;      -- Keep track of the number of
                                      -- characters
    end loop;
  end process;
end sample87;
```

VHDL 1076-1993 File Types

VHDL 1076-1993 allows files to be opened and closed as needed during simulation.

In VHDL '93, file types and associated functions and procedures were modified to allow files to be opened and closed as needed. In the 1987 specification, there is no provision for closing a file, and problems can arise when it is necessary for two parts of the same design description to open the same file at different points, or when existing files must be both read from and written to (as when appending data). The built-in functions available for file operations in VHDL '93 are:

File_open(f, fname, fmode)—Given a declared file object, file name (a string value) and a mode (either READ-MODE, WRITE_MODE, or APPEND_MODE), open the indicated file.

File_open(status, f, fname, fmode)—Same as above, but return the status of the file open request in the first parameter, which is of type **file_open_status**. The status returned is either OPEN_OK (meaning the file was successfully opened), STATUS_ERROR (meaning the file was not opened because there was already an open file associated with the file object), NAME_ERROR (meaning there was a system error related to the file name specified) or MODE_ERROR (meaning that the specified mode is not valid for the specified file).

File_close(f)—Close the specified file.

Read(f, object)—Given a declared file and an object, read one field of data from the file into that object. When the read procedure is invoked, data is read from the file and the file is advanced to the start of the next data field in the file.

Write(f, object)—Given a declared file and an object, write the data contained in the object to the file.

Endfile(f)—Given a declared file, return a boolean true value if the current file marker is at the end of the file.

A complete architecture that opens a file and loops through it, reading each character in the file, is shown below:

```
architecture sample93 of readfile is
begin
   Read_input: process
      type character_file is file of character;
      file cfile: character_file;
      variable C: character;
      variable char_cnt: integer := 0;
   begin
      file_open(cfile, "TESTFILE.ASC", READ_MODE);
      while not endfile(cfile) loop
         read (cfile, C) ;           -- Get a character from cfile into C
         char_cnt = char_cnt + 1;    -- Keep track of the number of
                                     -- characters
      end loop;
      file_close(cfile);
   end process;
end sample93;
```

Operators

VHDL is a rich language that includes many different operators.

The following charts summarize the operators available in VHDL. As indicated, not all operators can be used for all data types, and the data type that results from an operation may differ from the type of the object on which the operation is performed.

Note:
Operations defined for types Bit are also valid for type std_ulogic and std_logic.

Logical Operators

The logical operators and, or, nand, nor, xor and xnor are used to describe Boolean logic operations, or perform bit-wise operations, on bits or arrays of bits.

Operator	Description	Operand Types	Result Type
and	And	Any Bit or Boolean type	Same type
or	Or	Any Bit or Boolean type	Same type
nand	Not And	Any Bit or Boolean type	Same type
nor	Not Or	Any Bit or Boolean type	Same type
xor	Exclusive Or	Any Bit or Boolean type	Same type
nxor	Not Exclusive Or	Any Bit or Boolean type	Same type

Relational Operators

Relational operators are used to test the relative values of two scalar types. The result of a relational operation is always a Boolean true or false value.

Operator	Description	Operand Types	Result Type
=	Equality	Any type	Boolean
/=	Inequality	Any type	Boolean

<	Ordering	Any scalar type	Boolean
<=		or	
>		discrete array	
>=		type	

Adding Operators

The adding operators can be used to describe arithmetic functions or, in the case of array types, concatenation operations.

Operator	Description	Operand Types	Result Type
+	Addition	Any numeric type	Same type
-	Subtraction	Any numeric type	Same type
&	Concatenation	Any numeric type	Same type
&	Concatenation	Any array or element type	Same array type

Multiplying Operators

The multiplying operators can be used to describe mathematical functions on numeric types.

Note:
Synthesis tools vary in their support for multiplying operators.

Operator	Description	Operand Types	Result Type
*	Multiplication	Left: Any integer or floating point type Right: Same type	Same type
*	Multiplication	Left: Any physical type Right: Integer or real type	Same as left

*	Multiplication	Left: Any integer or real type Right: Any physical type	Same as right
/	Division	Any integer or floating point type	Same type
/	Division	Left: Any physical type Right: Any integer or real type	Same as left
/	Division	Left: Any physical type Right: Same type	Integer

Operator	Description	Operand Types	Result Type
mod	Modulus	Any integer type	Same type
rem	Remainder	Any integer type	Same type

Sign Operators

Sign operators can be used to specify the sign (either postive or negative) of a numeric object or literal.

Operator	Description	Operand Types	Result Type
+	Identity	Any numeric type	Same type
-	Negation	Any numeric type	Same type

Miscellaneous Operators

The exponentiation and absolute value operators can be applied to numeric types, in which case they result in the same numeric type. The logical negation operator results in the same type (bit or Boolean), but with the reverse logical polarity.

Operator	Description	Operand Types	Result Type
**	Exponentiation	Left: Any integer type Right: Integer type	Same as Left
**	Exponentiation	Left: Any floating point type Right: Integer type	Same as Left
abs	Absolute value	Any numeric type	Same numeric type
not	Logical negation	Any Bit or Boolean type	Same type

Shift Operators (1076-1993 only)

The shift operators (defined in 1076-1993) provide bit-wise shift and rotate operatons for arrays of type bit or Boolean.

Operator	Description	Operand Types	Result Type
sll	Shift left logical	Left: Any one-dimensinal array type whose element type is Bit or Boolean Right: Integer type	Same as left operand type
srl	Shift right logical	(same as above)	(same as above)
sla	Shift left arithmetic	(same as above)	(same as above)
sra	Shift right arithmetic	(same as above)	(same as above)
rol	Rotate left logical	(same as above)	(same as above)
ror	Rotate right logical	(same as above)	(same as above)

Attributes

Attributes provide additional information about various types of VHDL objects.

Attributes are a feature of VHDL that allow you to extract additional information about an object (such as a signal, variable or type) that may not be directly related to the value that the object carries. Attributes also allow you to assign additional information (such as data related to synthesis) to objects in your design description.

There are two classes of attributes: those that are predefined as a part of the 1076 standard, and those that have been introduced outside of the standard, either by you or by your design tool supplier.

Predefined Attributes

The VHDL specification describes five fundamental *kinds* of attributes. These five kinds of attributes are categorized by the results that are returned when they are used. The possible results returned from these attributes are: a value, a function, a signal, a type or a range.

Predefined attributes are always applied to a prefix...

Predefined attributes are always applied to a *prefix* (such as a signal or variable name, or a type or subtype name), as in the statement:

> **wait until** Clk = '1' **and** Clk'**event and** Clk'**last_value** = '0';

In this statement, the attributes **'event** and **'last_value** have been applied to the prefix **Clk**, which is a signal.

Some attributes also include parameters, so they are written in much the same way you would write a call to a function:

> **variable** V: state_type := state_type'**val**(2);   -- V has the value of
>   -- Strobe

In this case, the attribute **'val** has been applied to the prefix **state_type** (which is a type name) and has been given an attribute parameter, the integer value 2.

Value Kind Attributes: 'Left, 'Right, 'High, 'Low, 'Length, 'Ascending

The value kind attributes that return an explicit value and are applied to a type or subtype include the following:

'Left—This attribute returns the left-most element index (the *bound*) of a given type or subtype.

> Example:
>   **type** bit_array **is** array (1 **to** 5) **of** bit;
>   **variable** L: integer := bit_array**'left**;
>   -- L has a value of 1

'Right—Returns the right-most bound of a given type or subtype.

> Example:
>   **type** bit_array **is** array (1 **to** 5) **of** bit;
>   **variable** R: integer := bit_array**'right**;
>   -- R has a value of 5

'High—returns the upper bound of a given scalar type or subtype.

> Example:
>   **type** bit_array **is** array(-15 **to** +15) **of** bit;
>   **variable** H: integer := bit_array**'high**;
>   -- H has a value of 15

'Low—returns the upper bound of a given scalar type or subtype.

> Example:
>   **type** bit_array **is** array(15 **downto** 0) **of** bit;
>   **variable** L: integer := bit_array**'low**;
>   -- L has a value of 0

'Length—returns the length (number of elements) of an array.

> Example:
>   **type** bit_array **is** array (0 **to** 31) **of** bit;
>   **variable** LEN: integer := bit_array**'length**
>   -- LEN has a value of 32

'Ascending—(VHDL '93 attribute) returns a boolean true value of the type or subtype is declared with an ascending range.

Example:
```
type asc_array is array (0 to 31) of bit;
type desc_array is array (36 downto 4) of bit;
variable A1: boolean := asc_array'ascending;
    -- A1 has a value of true
variable A2: boolean := desc_array'ascending;
    -- A2 has a value of false
```

As you can see from the examples, value kind attributes (and all other predefined attributes) are identified by the ' (single quote) character (which is also called a *tick*). They are applied to type names, signals names and other identifiers, depending on the nature of the attribute. The value type attributes are used to determine the upper and lower (or left and right) bounds of a given type.

The following sample architecture uses the **'right** and **'left** attibutes to determine the left- and right-most element indices of an array in order to describe a width-independent shift operation:

```
architecture behavior of shifter is
begin
   reg: process(Rst,Clk)
   begin
     if Rst = '1' then   -- Async reset
        Qreg := (others => '0');
     elsif rising_edge(Clk) then
        Qreg := Data(Data'left+1 to Data'right) & Data(Data'left);
     end if;
   end process;
end behavior;
```

The **'right, 'left, 'high** and **'low** attributes can be used to return non-numeric values. The following example demonstrates how you can use the **'left** and **'right** attributes to identify the first and last items in an enumerated type:

```
architecture example of enums is
   type state_type is (Init, Hold, Strobe, Read, Idle);
   signal L, R: state_type;
begin
   L <= state_type'left;    -- L has the value of Init
   R <= state_type'right;   -- R has the value of Idle
end example;
```

Value Kind Attributes: 'Structure, 'Behavior

There are two additional value kind attributes that can be used to determine information about blocks or attributes in a design. These attributes, **'structure** and **'behavior**, return **true** or **false** values depending on whether the block or architecture being referenced includes references to lower-level components. The **'structure** attribute returns **true** if there are references to lower-level components, and **false** if there are no references to lower-level components. The **'behavior** attribute returns **false** if there are references to lower-level components, and **true** if there are no references to lower-level components.

'Structure—returns a **true** value if the prefix (which must be an architecture name) includes references to lower-level components.

'Behavior—returns a true value if the prefix (which must be an architecture name) does not include references to lower-level components.

Value Kind Attributes: 'Simple_name, 'Instance_name, 'Path_name

VHDL '93 adds three attributes that can be used to determine the precise configuration of entities in a design description. These attributes return information about *named entities*, which are various items that become associated with identifiers, character literals or operator symbols as the result of a declaration. For more information about these attributes, refer to the IEEE VHDL Language Reference Manual.

'Simple_name—(VHDL '93 attribute) returns a string value corresponding to the prefix, which must be a named entity.

'**Instance_name**—(VHDL '93 attribute) returns a string value corresponding to the complete path (from the design hierarchy root) to the named entity specified in the prefix, including the names of all instantiated design entities. The string returned by this attribute has a fixed format that is defined in the IEEE VHDL Language Reference Manual.

'**Path_name**—(VHDL '93 attribute) returns a string value corresponding to the complete path (from the design hierarchy root) to the named entity specified in the prefix. The string returned by this attribute has a fixed format that is defined in the IEEE VHDL Language Reference Manual.

Function Kind Attributes: 'Pos, 'Val, 'Succ, 'Pred, 'Leftof, 'Rightof

Attributes that return information about a given type, signal, or array value are called *function kind* attributes. VHDL defines the following function kind attributes that can be applied to types:

'**Pos(value)**—returns the position number of a type value.

> Example: **type** state_type **is** (Init, Hold, Strobe, Read, Idle);
> **variable** P: integer := state_type'**pos**(Read);
> -- P has the value of 3

'**Val(value)**—returns the value corresponding to a position number of a type value.

> Example: **type** state_type **is** (Init, Hold, Strobe, Read, Idle);
> **variable** V: state_type := state_type'**val**(2);
> -- V has the value of Strobe

'**Succ(value)**—returns the value corresponding to position number after a given type value.

> Example: **type** state_type **is** (Init, Hold, Strobe, Read, Idle);
> **variable** V: state_type := state_type'**succ**(Init);
> -- V has the value of Hold

'Pred(value)—returns the value corresponding to position number preceding a given type value.

Example: **type** state_type **is** (Init, Hold, Strobe, Read, Idle);
             variable V: state_type := state_type'**pred**(Hold);
                          -- V has the value of Init

'Leftof(value)—returns the value corresponding to position number to the left of a given type value.

Example: **type** state_type **is** (Init, Hold, Strobe, Read, Idle);
             variable V: state_type := state_type'**leftof**(Idle);
                          -- V has the value of Read

'Rightof(value)—returns the value corresponding to position number to the right of a given type value.

Example: **type** state_type **is** (Init, Hold, Strobe, Read, Idle);
             variable V: state_type := state_type'**rightof**(Read);
                          -- V has the value of Idle

From the above descriptions, it might appear that the **'val** and **'succ** attributes are equivalent to the attributes **'leftof** and **'rightof**. One case where they would be different is the case where a subtype is defined that changes the ordering of the base type:

```
type state_type is (Init, Hold, Strobe, Read, Idle);
subtype reverse_state_type is state_type range Idle downto Init;

variable V1: reverse_state_type := reverse_state_type'leftof(Hold);
             -- V1 has the value of Strobe
variable V2: reverse_state_type := reverse_state_type'pred(Hold);
             -- V2 has the value of Init
```

Function Kind Array Attributes: 'Left, 'Right, 'High, 'Low

The function kind attributes that can be applied to array objects include:

'Left(value)—returns the index value corresponding to the left bound of a given array range.

Example: **type** bit_array **is** array (15 downto 0) of bit;
            variable I: integer :=
                bit_array**'left**(bit_array**'range**);
                -- I has the value of 15

'Right(value)—returns the index value corresponding to the right bound of a given array range.

Example: **type** bit_array **is** array (15 downto 0) of bit;
            variable I: integer :=
                bit_array**'right**(bit_array**'range**);
                -- I has the value of 0

'High(value)—returns the index value corresponding to the upper-most bound of a given array range.

Example: **type** bit_array **is** array (15 downto 0) of bit;
            variable I: integer :=
                bit_array**'high**(bit_array**'range**);
                -- I has the value of 15

'Low(value)—returns the index value corresponding to the lower bound of a given array range.

Example: **type** bit_array **is** array (15 downto 0) of bit;
            variable I: integer :=
                bit_array**'low**(bit_array**'range**);
                -- I has the value of 0

Function Kind Attributes: 'Event, 'Active, 'Last_event, 'Last_value, 'Last_active

Function kind attributes that return information about signals (such as whether that signal has changed its value or its previous value) include:

'Event—returns a true value of the signal had an event (changed its value) in the current simulation delta cycle.

Example: **process**(Rst,Clk)
            begin
                if Rst = '1' **then**
                    Q <= '0';

103

```
    elsif Clk = '1' and Clk'event then
                -- Look for clock edge
        Q <= D;
    end if;
end process;
```

'Active—returns true if any transaction (scheduled event) occurred on this signal in the current simulation delta cycle.

Example:
```
process
variable A,E: boolean;
begin
    Q <= D after 10 ns;
    A := Q'active;  -- A gets a value of True
    E := Q'event;   -- E gets a value of False
    . . .
end process;
```

'Last_event—returns the time elapsed since the previous event occurring on this signal.

Example:
```
process
variable T: time;
begin
    Q <= D after 5 ns;
    wait 10 ns;
    T := Q'last_event;    -- T gets a value of 5 ns
    . . .
end process;
```

'Last_value—returns the value of the signal prior to the last event.

Example:
```
process
variable V: bit;
begin
    Q <= '1';
    wait 10 ns;
    Q <= '0';
    wait 10 ns;
    V := Q'last_value;     -- V gets a value of  '1'
    . . .
end process;
```

'Last_active—returns the time elapsed since the last transaction (scheduled event) of the signal.

Example:
```
process
variable T: time;
begin
    Q <= D after 30 ns;
    wait 10 ns;
    T := Q'last_active;   -- T gets a value of 10 ns
    . . .
end process;
```

Note:

The 'active, 'last_event, 'last_value and 'last_active attributes are not generally supported in synthesis tools. Of the preceding five attributes, only 'event should be used when describing synthesizable registered circuits. The 'active, 'last_event, 'last_value and 'last_active attributes should only be used to describe circuits for test purposes (such as for setup and hold checking). If they are encountered by a synthesis program, they will either be ignored, or the program will return an error and halt operation.

Function Kind Attributes: 'Image, 'Value

The **'image** and **'value** attributes were added in the 1993 specification to simplify the reporting of information through Text I/O. These attributes both return string results corresponding to their parameter values.

'Image(expression)—(VHDL '93 attribute) returns a string representation of the expression parameter, which must be of a type corresponding to the attribute prefix.

Example:
```
assert (Data.Q = '1')
    report "Test failed on vector " &
            integer'image(vector_idx)
    severity WARNING;
```

'Value(string)—(VHDL '93 attribute) returns a value, of a type specified by the prefix, corresponding to the parameter string.

Example:
    write(a_outbuf,string'("Enter desired state (example: S1)"));
    writeline(OUTPUT,a_outbuf);
    readline(INPUT,a_inbuf);
    read(a_inbuf,instate); -- instate is a string type

    next_state <= state_type'**value**(instate);
                  -- convert string to type state_type

    write(a_outbuf,string'("Enter duration (example: 15)"));
    writeline(OUTPUT,a_outbuf);
    readline(INPUT,a_inbuf);
    read(a_inbuf,induration);   -- induration is a string type

    duration <= integer'**value**(induration);
                  -- convert string to type integer

Signal Kind Attributes: 'Delayed, 'Stable, 'Quiet, 'Transaction

The signal kind attributes are attributes that, when invoked, create special signals that have values and types based on other signals. These special signals can then be used anywhere in the design description that a normally declared signal could be used. One example of where you might use such an attribute is to create a series of delayed clock signals that are all based on the waveform of a base clock signal.

Signal kind attributes include the following:

    'Delayed(time)—creates a delayed signal that is identical in waveform to the signal the attribute is applied to. (The time parameter is optional, and may be omitted.)

Example: **process**(Clk'**delayed**(hold))
                      -- Hold time check for input Data
        begin
          if Clk = '1' and Clk'**stable**(hold) **then**
             assert(Data'stable(hold))
                report "Data input failed hold time check!"
                severity warning;

```
        end if;
      end process;
```

'**Stable (time)**—creates a signal of type **boolean** that
  becomes **true** when the signal is stable (has no event)
  for some given period of time.

Example:
```
process
variable A: Boolean;
begin
    wait for 30 ns;
    Q <= D after 30 ns;
    wait 10 ns;
    A := Q'stable(20 ns);
        -- A gets a value of true (event has not
        -- yet occurred)
    wait 30 ns;
    A := Q'stable(20 ns);
        -- A gets a value of false (only 10 ns
        -- since event)
    . . .
end process;
```

'**Quiet (time)**—creates a signal of type **boolean** that becomes true when the signal has no transactions (scheduled events) or actual events for some given period of
  time.

Example:
```
process
variable A: Boolean;
begin
    wait for 30 ns;
    Q <= D after 30 ns;
    wait 10 ns;
    A := Q'quiet(20 ns);
        -- A gets a value of false (10 ns since
        -- transaction)
    wait 40 ns;
    A := Q'quiet(20 ns);
        -- A finally gets a value of true (20 ns
        -- since event)
    . . .
end process;
```

'**Transaction**—creates a signal of type **bit** that toggles its value whenever a transaction or actual event occurs on the signal the attribute is applied to.

Type Kind Attribute: 'Base

'**Base**—returns the base type for a given type or subtype.

Example:

```
type mlv7 is ('0','1','X','Z','H','L','W');
subtype mlv4 is mlv7 range '0' to 'Z';
variable V1: mlv4 := mlv4'right;
    -- V1 has the value of 'Z'
variable V2: mlv7 := mlv4'base'right;
    -- V2 has the value of 'W'
variable I1: integer := mlv4'width;
    -- I1 has the value of 4
variable I2: integer := mlv4'base'width;
    -- I2 has the value of 7
```

Range Kind Attributes: 'Range, 'Reverse_range

The range kind attributes return a special value that is a range, such as you might use in a declaration or looping scheme.

'**Range**—returns the range value for a constrained array.

Example:

```
function parity(D: std_logic_vector) return
        std_logic is
    variable result: std_logic := '0';
begin
    for i in D'range loop
        result := result xor D(i);
    end loop;
    return result;
end parity;
```

'**Reverse_range**—returns the reverse of the range value for a constrained array.

Example:

```
STRIPX: for i in D'reverse_range loop
        if D(i) = 'X' then
            D(i) = '0';
```

```
        else
            exit;        -- only strip the terminating Xs
        end if;
    end loop;
```

Custom Attributes

Custom attributes can be used to attach information to (or decorate) a VHDL object.

Custom attributes are those attributes that are not defined in the IEEE specifications, but that you (or your simulation or synthesis tool vendor) define for your own use. A good example is the attribute **enum_encoding**, which is provided by a number of synthesis tool vendors (most notably Synopsys) to allow specific binary encodings to be attached to objects of enumerated types.

An attribute such as **enum_encoding** is declared (again, either by you or by your design tool vendor) using the following method:

```
    attribute enum_encoding: string;
```

This attribute could be written directly in your VHDL design description, or it could have been provided to you by the tool vendor in the form of a package. Once the attribute has been declared and given a name, it can be referenced as needed in the design description:

```
    type statevalue is (INIT, IDLE, READ, WRITE, ERROR);
    attribute enum_encoding of statevalue: type is "000 001 011 010 110";
```

When these declarations are processed by a synthesis tool that supports the **enum_encoding** attribute, information about the encoding of the type **statevalue** will be used by that tool. When the design is processed by design tools (such as simulators) that do not recongnize the **enum_encoding** attribute, it will simply be ignored.

Custom attributes are a convenient "back door" feature of VHDL, and design tool vendors have created many such attributes to give you more control over the synthesis and simulation process. For detailed information about custom attributes, refer to your design tool documentation.

Type Conversions and Type Marks

VHDL is a strongly typed language...

VHDL is a strongly typed language, meaning that you cannot simply assign a literal value or object of one type to an object of another type. To allow the transfer of data between objects of different types, VHDL includes *type conversion* features for types that are closely related. VHDL also allows type conversion functions to be written for types that are not closely related. In addition, VHDL includes type mark features to help specify (or *qualify*) the type of a literal value when the context or format of the literal makes its type ambiguous.

Explicit Type Conversions

Explicit type conversions are allowed between closely-related types.

The simplest type conversions are *explicit* type conversions, which are only allowed between *closely related* types. Two types are said to be closely related when they are either abstract numeric types (integers or floating points), or if they are array types of the same dimensions and share the same types (or the element types themselves are closely related) for all elements in the array. In the case of two arrays, it is not necessary for the arrays to have the same direction. If two subtypes share the same base type, then no explicit type conversion is required.

The following example demonstrates implicit and explicit type conversions:

```
architecture example of typeconv is
    type array1 is array(0 to 7) of std_logic;
    type array2 is array(7 downto 0) of std_logic;
    subtype array3 is std_logic_vector(0 to 7);
    subtype array4 is std_logic_vector(7 downto 0);
    signal a1: array1;
    signal a2: array2;
```

```
        signal a3: array3;
        signal a4: array4;

    begin
        a2 <= array2(a1);   -- explicit type conversion
        a4 <= a3;           -- no explicit type conversion needed
    end example;
```

Type Conversion Functions

Type conversion functions can be used to assign a value of one type to an object of a different type.

To convert data from one type to an unrelated type (such as from an integer type to an array type), you must make use of a type conversion function. Type conversion functions may be obtained from standard libraries (such as the IEEE 1164 library), from vendor-specific libraries (such as those supplied by synthesis tool vendors), or you can write you own type conversion functions.

A type conversion function is a function that accepts one argument of a specified type and returns the equivalent value in another type. The following two functions are examples of type conversion functions that convert between **integer** and array (**std_ulogic_vector**) data types:

```
    ---------------------------------------------------------------------
    -- Convert a std_ulogic_vector to an unsigned integer
    --
    function to_uint (a: std_ulogic_vector) return integer is
        alias av: std_ulogic_vector (1 to a'length) is a;
        variable val: integer := 0;
        variable b: integer := 1;
    begin
        for i in a'length downto 1 loop
            if (av(i) = '1') then    -- if LSB is '1',
                val := val + b;      -- add value for current bit position
            end if;
            b := b * 2;   -- Shift left 1 bit
        end loop;

        return val;
    end to_uint;
```

111

```
--------------------------------------------------------
-- Convert an integer to a std_ulogic_vector
--
function to_vector (size: integer; val: integer) return std_ulogic_vector is
   variable vec: std_ulogic_vector (1 to size);
   variable a: integer;
begin
   a := val;
   for i in size downto 1 loop
     if ((a mod 2) = 1) then
        vec(i) := '1';
     else
        vec(i) := '0';
     end if;
     a := a / 2;
   end loop;
   return vec;
end to_vector;
```

The following example (a loadable counter) demonstrates how these two functions could be used:

Type conversion functions are often required when describing arithmetic operations.

```
library ieee;
use ieee.std_logic_1164.all;

library types;      -- Type conversions have been compiled into library 'types'
use types.conversions.all;

entity count16 is
   port (Clk,Rst,Load: in std_ulogic;
        Data: In std_ulogic_vector(3 downto 0);
        Count: out std_ulogic_vector(3 downto 0));
end count16;

architecture count16a of count16 is
begin
   process(Rst,Clk)
     variable Q: integer range 0 to 15;
   begin
     if Rst = '1' then            -- Asynchronous reset
        Q := 0;
     elsif rising_edge(Clk) then
        if Load = '1' then
           Q := to_uint(Data);  -- Convert vector to integer
        elsif Q = 15 then
           Q := 0;
```

```
    else
        Q := Q + 1;
    end if;
  end if;

  Count <= to_vector(4,Q);        -- Convert integer to vector
                                  -- for use outside the process.
  end process;
end count16a;
```

In this example, the interface specified in the entity port list uses standard logic data types, including a **std_ulogic_vector** array data type for the counter output. Because there are no arithmetic operations defined for the **std_ulogic_vector** data type, it is necessary to introduce an intermediate integer variable and convert the **Data** input from a **std_ulogic_vector** type to an integer when assigning it to the intermediate variable, and to convert the intermediate variable back to a **std_ulogic_vector** array type when assigning it to the **Count** output.

Synthesis Note:

*The preceding example is unlikely to be synthesizable because the type conversion functions (**to_uint()** and **to_vector()**) are written using unconstrained integers. As a practical matter, you should never write an arbitrary-width type conversion function that you intend to use in a synthesizable design description. Instead, you should make use of type conversion functions provided by your synthesis vendor or use the 1076.3 **signed** or **unsigned** type (see Chapter 4, Using Standard Logic).*

Another common application of type conversion functions is the conversion of string data read from a file to array or record data types suitable for use as stimulus in a test bench. The following function accepts data in the form of a fixed-length string and converts it, character by character, into a record data type:

```
type test_record is record
  CE: std_ulogic; -- Clock enable
  Set: std_ulogic; -- Preset
  Din: std_ulogic; -- Binary data input
  Doutput: std_ulogic_vector (15 downto 0); -- Expected output
```

```
    end record;

    function str_to_record(s: string(18 downto 0)) return test_record is
       variable temp: test_record;
    begin
       case s(18) is
          when '1' => temp.CE := '1';
          when '0' => temp.CE := '0';
          when others => temp.CE = 'X';
       end case;
       case s(17) is
          when '1' => temp.Set := '1';
          when '0' => temp.Set := '0';
          when others => temp.Set = 'X';
       end case;
       case s(16) is
          when '1' => temp.Din := '1';
          when '0' => temp.Din := '0';
          when others => temp.Din = 'X';
       end case;
       for i in 15 downto 0 loop
          case s(i) is
             when '1' => temp.Doutput := '1';
             when '0' => temp.Doutput := '0';
             when others => temp.Doutput = 'X';
          end case;
       end loop;
       return temp;
    end str_to_record;
```

Rather than write your own type coversion functions, you should make use of standard or pre-written conversion functions whenever possible.

There are many applications of type conversion functions, and many possible ways to write them. If you are writing a synthesizable design description, you should (whenever possible) make use of type conversions that have been provided to you by your synthesis vendor, as type conversion functions can be difficult (in some cases impossible) for synthesis tools to handle.

Ambiguous Literal Types

Functions and procedures in VHDL are uniquely identified not only by their names, but also by the types of their arguments. (See Subprogram Overloading in Chapter 7, *Creating Modular Designs*.) This means that you can, for example, write two functions to perform similar tasks, but on different types of input data. The ability to overload functions and procedures can lead to ambiguities when functions are called, if the types of one or more arguments are not explicitly stated.

For example, consider two type conversion functions with the following interface declarations:

```
function to_integer (vec: bit_vector) return integer is
    . . .
end to_uint;

function to_integer (s: string) return integer is
    . . .
end to_uint;
```

If you were to write an assignment statement such as:

```
architecture ambiguous of my_entity is
    signal Int35: integer;
begin
    Int35 <= to_integer("00100011");   -- This will produce an error
    . . .
end ambiguous;
```

Literal values can sometimes be ambiguous in terms of their implied data types...

then the compiler would produce an error message because it would be unable to determine which of the two functions is appropriate—the literal "00100011" could be either a **string** or **bit_vector** data type.

To remove data type ambiguity in such cases, you have two options: you can either introduce an intermediate constant, signal or variable, as in:

```
architecture unambiguous1 of my_entity is
    constant Vec35: bit_vector := "00100011";
    signal Int35: integer;
begin
    Int35 <= to_integer(Vec35);
    . . .
end unambiguous1;
```

or introduce a type mark to qualify the argument, as in:

```
architecture unambiguous2 of my_entity is
    signal Int35: integer;
begin
    Int35 <= to_integer(bit_vector'"00100011");
    . . .
end unambiguous2;
```

Resolved and Unresolved Types

A signal requires *resolution* whenever it is simultaneously driven with more than one value. By default, data types (whether standard types or types you define) are unresolved, resulting in errors being generated when there are multiple values being driven onto signals of those types. These error messages may be the desired behavior, as it is usually a design error when such conditions occur. If you actually intend to drive a signal with multiple values (as in the case of a bus interface), then you will need to use a *resolved* data type.

Resolved data types allow multiple values to be driven onto a signal simultaneously with a known result.

Data types are resolved only when a *resolution function* has been included as a part of their definition. A resolution function is a function that specifies, for all possible combinations of one or more input values (expressed as an array of the data type being resolved), what the resulting (resolved) value will be.

The following sample package defines a resolved data type consisting of four possible values, '0', 1', 'X' and 'Z'. The resolution function covers all possible combinations of input values and specifies the resolved value corresponding to each combination:

```
package types is
    type xbit is ( '0',  -- Logical  0
                   '1',  -- Logical  1
                   'X',  -- Unknown
                   'Z'   -- High Impedance );

    -- unconstrained array is required for the resolution function...
    type xbit_vector is array ( natural range <> ) of xbit;

    -- resolution function...
```

```
function resolve_xbit ( v : xbit_vector ) return xbit;

-- resolved logic type...
subtype xbit_resolved is resolve_xbit xbit;

end types;

package body types is

-- Define resolutions as a table...

type xbit_table is array(xbit, xbit) of xbit;
constant resolution_table: xbit_table := (
--      0   1   X   Z
      ( '0', 'X', 'X', '0' ), -- 0
      ( 'X', '1', 'X', '1' ), -- 1
      ( 'X', 'X', 'X', 'X' ), -- X
      ( '0', '1', 'X', 'Z' ) -- Z
);


function resolve_xbit ( v: xbit_vector ) return xbit is
    variable result: xbit;
begin
    -- test for single driver
    if (v'length = 1) then
       result := v(v'low); -- Return the same value if only 1 value
    else
       result := 'Z';
       for i in v'range loop
          result := resolution_table(result, v(i));
       end loop;
    end if;
    return result;
end resolve_xbit;

end types;
```

The resolution function is invoked automatically whenever a signal of the associated type is driven with one or more values. The array argument **v** represents all of the values being driven onto the signal at any given time.

With the types **xbit** and **xbit_resolved** defined in this way, the resolved data type **xbit_resolved** can be used for situations in which resolutions are required. The following example shows how the resolved type **xbit_resolved** could be used to describe the operation of a pair of three-state signals driving a common signal:

A resolved data type could be used to describe the behavior of a three-state output.

```
use work.types.all;
entity threestate is
    port (en1, en2: in xbit_resolved;
          A,B: in xbit_resolved;
          O: out xbit_resolved);
end threestate;


architecture sample of threestate is
    signal tmp1,tmp2: xbit_resolved;
begin

    tmp1 <= A when en1 else 'Z';
    tmp2 <= B when en2 else 'Z';

    O <= tmp1;
    O <= tmp2;

end sample;
```

In this example, the output **O** could be driven with various combinations of the values of **A** and **B** and the value **'Z'**, depending on the states of the two inputs **en1** and **en2**. The resolution function takes care of calculating the correct value for **O** for any of these combinations during simulation.

More information about resolved types, in particular the standard resolved types **std_logic** and **std_logic_vector**, can be found in Chapter 4, *Using Standard Logic*.

Moving On

Now that we've seen the variety of building blocks available in VHDL, it's easy to imagine constructing both simple and complex designs.

But wait. There's more. We can make VHDL design even easier by...

4. Using Standard Logic

In this chapter, we'll take a close look at two important standards that augment Standard 1076, adding important capabilities for both simulation and synthesis. These two standards are IEEE Standards 1164 and 1076.3.

IEEE Standard 1164

IEEE Standard 1164 was released in the late 1980s, and helped to overcome an important limitation of VHDL and its various commercial implementations. These limitations were created by the fact that VHDL, while being rich in data types, did not include a standard type that would allow multiple values (high-impedance, unknown, etc.) to be represented for a wire. These *metalogic* values are important for accurate simulation, so VHDL simulation vendors were forced to invent their own proprietary data types using syntactically correct, but non-standard, enumerated types.

IEEE 1164 replaces proprietary data types...

IEEE 1164 replaces these proprietary data types (which include systems having four, seven, or even thirteen unique values) with a standard data type having nine values, as shown in Figure 4-1.

Value	Description
'U'	Uninitialized
'X'	Unknown
'0'	Logic 0 (driven)
'1'	Logic 1 (driven)
'Z'	High impedance
'W'	Weak 1
'L'	Logic 0 (read)
'H'	Logic 1 (read)
'-'	Don't-care

Figure 4-1. The standard logic data type includes nine distinct values to represent the possible states of a signal.

These nine values make it possible to accurately model the behavior of a digital circuit during simulation. For synthesis users, the standard has additional benefits for describing circuits that involve output enables, as well as for specifying don't-care logic that can be used to optimize the combinational logic requirements of a circuit.

Advantages of IEEE 1164

Standard logic data types increase simulation accuracy and improve design portability.

There are many compelling reasons to adopt IEEE Standard 1164 for all of your design efforts and to use it as a standard data type for all system interfaces. For simulation purposes, the standard logic data types allow you to apply values other than '0' or '1' as inputs and view the results. This capability could be used, for example, to verify that an input with an unknown (uninitialized or don't-care) value does not cause the circuit to behave in an unexpected manner. The *resolved* standard logic data types can be used to model the behavior of multiple drivers in your circuit. You might use these types to model, for example, the behavior of a three-state bus driver.

The most important reason to use standard logic data types is portability: if you will be interfacing to other components during simulation (such as those obtained from third party simulation model providers) or moving your design description between different simulation environments, then IEEE 1164 gives you a standard, portable style with which to describe your circuit.

Using The Standard Logic Package

To use the IEEE 1164 standard logic data types, you will need to add at least two statements to your VHDL source files. These statements (shown below) cause the IEEE 1164 standard library (named **ieee**) to be loaded and its contents (the **std_logic_1164** package) made visible:

*Place **library** and **use** statements prior to all design units referencing standard logic data types.*

```
library ieee;
use ieee.std_logic_1164.all;
```

In most design descriptions, you will place these two statements at the top of your source file, and repeat them as needed prior to subsequent design units (entity and architecture pairs) in the file. If your source file includes more than one design unit, you need to repeat the **use** statement just prior to each design unit in order to make the contents of the standard library visible to each design unit, as shown below:

```
library ieee;
use ieee.std_logic_1164.all;
package my_package is

   . . .
end my_package;


use ieee.std_logic_1164.all;
entity first_one is

   . . .
end first_one;


use ieee.std_logic_1164.all;
architecture structure of first_one is

   . . .
end structure;


use ieee.std_logic_1164.all;
entity second_one is
```

123

...
end second_one;

Note:
*VHDL has special visibility rules for architectures: it is not necessary to place a **use** statement prior to an architecture declaration if the corresponding entity declaration includes a **use** statement. In the above example, the **use** statement appearing prior to the architecture **structure** is not actually needed and could be omitted.*

Once you have included the **ieee** library and made the **std_logic_1164** package visible in your design description, you can make use of the data types, operators and functions provided for you as a part of the standard.

Standard logic data types are provided in resolved and unresolved forms.

There are two fundamental data types provided for you in the **std_logic_1164** package. These data types, **std_logic** and **std_ulogic**, are enumerated types defined with nine symbolic (single character) values. The following definition of **std_ulogic** is taken directly from the IEEE 1164 standard:

```
type std_ulogic is ( 'U',  -- Uninitialized
                     'X',  -- Forcing  Unknown
                     '0',  -- Forcing  0
                     '1',  -- Forcing  1
                     'Z',  -- High Impedance
                     'W',  -- Weak     Unknown
                     'L',  -- Weak     0
                     'H',  -- Weak     1
                     '-'   -- Don't care
                   );
```

The **std_ulogic** data type is an *unresolved* type, meaning that it is illegal for two values (such as **'0'** and **'1'**, or **'1'** and **'Z'**) to be simultaneously driven onto a signal of type **std_ulogic**. If you are not describing a circuit that will be driving different values onto a wire (as you might in the case of a bus interface), then you might want to use the **std_ulogic** data type to help catch errors (such as incorrectly specified, overlapping combinational logic) in your design description. If you are describing a circuit that involves multiple values being driven onto a wire, then you will need to use the type **std_logic**. **Std_logic** is a resolved type based on **std_ulogic**. Resolved types are declared with resolution functions, as described in Chapter 3, *Exploring Objects and Data Types*. Resolution functions define the resulting behavior when an object is driven with multiple values simultaneously.

*Use standard logic data types as a replacement for **bit** and **bit_vector** data types.*

When using either of these data types, you will use them as one-for-one replacements for the built-in type **bit**. The following example shows how you might use the **std_logic** data type to describe a simple NAND gate coupled to an output enable:

```
library ieee;
use ieee.std_logic_1164.all;
entity nandgate is
    port (A, B, OE: in std_logic; Y: out std_logic);
end nandgate;

architecture arch1 of nandgate is
    signal n: std_logic;
begin
    n <= not (A and B);
    Y <= n when OE = '0' else 'Z';
end arch1;
```

Note:
*As written, it is not actually necessary for this circuit to be described using the resolved type **std_logic** for correct simulation. Operated as a stand-alone circuit, the output **Y** will never be driven with two different values. When connected through hierarchy into a larger circuit, however, it is highly likely that such a situation will occur, and **std_logic** will thus be required.*

Std_logic_vector and Std_ulogic_vector

In addition to the single-bit data types **std_logic** and **std_ulogic**, IEEE Standard 1164 includes array types corresponding to each of these types. Both **std_logic_vector** and **std_ulogic_vector** are defined in the **std_logic_1164** package as unbounded arrays similar to the built-in type **bit_vector**. In practice, you will probably use **std_logic_vector** or **std_ulogic_vector** with an explicit width, or you will use a subtype to create a new data type based on **std_logic_vector** or **std_ulogic_vector** of the width required. The following sample design description uses a subtype (defined in an external package) to create an 8-bit array based on **std_ulogic_vector**:

```
library ieee;
use ieee.std_logic_1164.all;
```

Use a subtype to create a constrained version of std_ulogic_vector or std_logic_vector.

```
package my_types is
    subtype std_byte is std_ulogic_vector(7 downto 0);
end my_types;
```

```
use ieee.std_logic_1164.all;
entity shiftl is
    port (DataIN: in std_byte; DataOUT: out std_byte; Err: out std_ulogic);
end shiftl;
```

```
architecture arch1 of shiftl is
    signal n: std_logic;
begin
    DataOUT <= DataIN(DataIN'left - 1 downto 0) & '0';    -- Shift left one bit
    Err <= DataIN(DataIN'left);    -- Check for overflow
end arch1;
```

In this example (an 8-bit shifter), the subtype **std_byte** is defined in terms of **std_ulogic_vector** and can be used to replace **std_ulogic_vector(7 downto 0)** throughout the design description. The circuit is described in such a way that the width of the shifter is dependent only on the width of the type **std_byte**, so it is easy to modify the width of the circuit later.

Type Conversion and Standard Logic

If you need to describe operations such as counters that are not directly supported in the standard logic data types, you will almost certainly have to make use of type conversion functions to convert the standard logic data types at your system interfaces to types such as integers that support such operations.

As described in Chapter 3, *Exploring Objects and Data Types*, type conversion functions are functions that accept an object of one data type and return the equivalent data value represented as a different data type. Some type conversion functions are provided in the IEEE 1164 **std_logic_1164** package (functions to convert between **std_logic_vector** and **bit_vector**, for example), but no functions are provided in that package to convert between standard logic data types and numeric data types such as integers.

Arithmetic operations are not directly supported in IEEE 1164, so type conversions may be required.

Arithmetic circuits (such as adders and counters) are common elements of modern digital systems, and of design descriptions intended for synthesis. So what do you do if you want to use standard logic data types *and* describe arithmetic operations? There are actually a number of possible solutions to this problem.

The first solution is to write your own synthesizable type conversion functions, so that you can translate between standard logic values that you will use for your system interfaces (such as the ports for your entities) and the internal numeric type signals and variables you will need to describe your arithmetic function. This is actually a rather poor solution, as it can be quite difficult (perhaps impossible) to write a general-purpose (meaning width-independent) type conversion function that your synthesis tool can handle.

The second solution is to make use of custom type conversion functions or data types that have been provided by your synthesis vendor for use with their tool. An example of such a method (using the **std_logic_arith** package provided by Synopsys) is shown below:

```
library ieee;
use  ieee.std_logic_1164.all;
use  ieee.std_logic_arith.all;

entity COUNT16 is
    port (Clk,Rst,Load: in std_logic;
        Data: in std_logic_vector(3 downto 0);
        Count: out std_logic_vector(3 downto 0)
    );
end COUNT16;
```

127

```vhdl
architecture COUNT16_A of COUNT16 is
begin
  process(Rst,Clk)
      -- The unsigned integer type is defined in synopsys.vhd...
      variable Q: unsigned (3 downto 0);
  begin
    if Rst = '1' then
      Q := "0000";
    elsif rising_edge(Clk) then
      if Load = '1' then
        for i in 3 downto 0 loop
          Q(i) := Data(i);
        end loop;
      elsif Q = "1111" then
        Q := "0000";
      else
        Q := Q + "0001";
      end if;
    end if;

    Count <= conv_std_logic_vector(Q,Data'length);

  end process;

end COUNT16_A;
```

*Using a specially-defined numeric type such as **unsigned** can make counters and arithmetic circuits easier to describe.*

In this example, the **conv_std_logic_vector** function has been provided in the **std_logic_arith** package, which was supplied by a synthesis vendor (in this case, Synopsys).

Using synthesis tool-specific packages such as **std_logic_arith** can be quite convenient, but may result in a non-portable design description. (This is particularly true if you use tool-specific type conversion functions, which often have completely different naming conventions and function parameters, and are typically incompatible with synthesis tools other than those they were originally written for.)

The best solution to the numeric data is to use the IEEE 1076.3 standard numeric data types, which are described later in this chapter.

Standard Logic Data Types

This section describes in detail the contents of the IEEE 1164 Standard Logic package **std_logic_1164**. The **std_logic_1164** package is compiled into a library named **ieee**, and includes the following data type and function definitions:

Type Std_ulogic

Type std_ulogic is intended to represent a single wire that can have various logical (and metalogical) values. **Std_ulogic** is the base type for other IEEE 1164 (and related) standard types, including **std_logic**, **std_logic_vector**, **signed** and **unsigned**. **Std_ulogic** has the following definition:

```
type std_ulogic is ( 'U',  -- Uninitialized
                     'X', -- Forcing  Unknown
                     '0', -- Forcing  0
                     '1', -- Forcing  1
                     'Z', -- High Impedance
                     'W', -- Weak     Unknown
                     'L', -- Weak     0
                     'H', -- Weak     1
                     '-'  -- Don't care
                   );
```

The **std_ulogic** data type is an enumerated type similar is usage to the **bit** data type provided in the standard (1076) library. **Std_ulogic** is an unresolved type.

Type Std_ulogic_vector

Type **std_ulogic_vector** is intended to represent a collection of wires, or a bus of arbitrary width. **Std_ulogic_vector** has the following definition:

```
type std_ulogic_vector is array ( natural range <> ) of std_ulogic;
```

Std_ulogic_vector is an unconstrained array of **std_ulogic**, and is analogous to the standard type **bit_vector**.

129

Type Std_logic

Type **std_logic** is a resolved type based on **std_ulogic**, and has the following definition:

subtype std_logic **is resolved** std_ulogic;

In the case of multiple drivers, the nine values of **std_logic** are resolved to values as indicated in the chart of Figure 4-2.

Figure 4-2. The ***std_logic*** *data type is a resolved type, with multiple values resolved to a single (resolved) value.*

	'U'	'X'	'0'	'1'	'Z'	'W'	'L'	'H'	'-'
'U'	'U'	'U'	'U'	'U'	'U'	'U'	'U'	'U'	'U'
'X'	'U'	'X'	'X'	'X'	'X'	'X'	'X'	'X'	'X'
'0'	'U'	'X'	'0'	'X'	'0'	'0'	'0'	'0'	'X'
'1'	'U'	'X'	'X'	'1'	'1'	'1'	'1'	'1'	'X'
'Z'	'U'	'X'	'0'	'1'	'Z'	'W'	'L'	'H'	'X'
'W'	'U'	'X'	'0'	'1'	'W'	'W'	'W'	'W'	'X'
'L'	'U'	'X'	'0'	'1'	'L'	'W'	'L'	'W'	'X'
'H'	'U'	'X'	'0'	'1'	'H'	'W'	'W'	'H'	'X'
'-'	'U'	'X'	'X'	'X'	'X'	'X'	'X'	'X'	'X'

Type Std_logic_vector

type std_logic_vector **is** array (natural **range** <>) **of** std_logic;

Std_logic_vector is an unconstrained array of **std_logic**.

Subtypes Based on Std_ulogic

subtype X01 **is resolved** std_ulogic **range** 'X' **to** '1'; -- ('X','0','1')
subtype X01Z **is resolved** std_ulogic **range** 'X' **to** 'Z'; -- ('X','0','1','Z')
subtype UX01 **is resolved** std_ulogic **range** 'U' **to** '1'; -- ('U','X','0','1')
subtype UX01Z **is resolved** std_ulogic **range** 'U' **to** 'Z'; -- ('U','X','0','1','Z')

The **X01**, **X01Z**, **UX01**, and **UX01Z** subtypes are used within the **std_logic_1164** package to simplify various operations on standard logic data, and may also be used when you have a need for 3-, 4-, or 5-valued logic systems.

Standard Logic Operators

The following operators are defined for types **std_ulogic**, **std_logic**, **std_ulogic_vector** and **std_logic_vector**:

Logical Operators

function "and" (l : std_ulogic; r : std_ulogic) **return** UX01;
function "nand" (l : std_ulogic; r : std_ulogic) **return** UX01;
function "or"   (l : std_ulogic; r : std_ulogic) **return** UX01;
function "nor" (l : std_ulogic; r : std_ulogic) **return** UX01;
function "xor" (l : std_ulogic; r : std_ulogic) **return** UX01;
function "xnor" (l : std_ulogic; r : std_ulogic) **return** ux01;

*Note: the **xnor** operator is supported in standard 1076-1993 only.*

function "not" (l : std_ulogic) **return** UX01;

Array Logical Operators

function "and" (l, r : std_logic_vector) **return** std_logic_vector;
function "and" (l, r : std_ulogic_vector) **return** std_ulogic_vector;
function "nand" (l, r : std_logic_vector) **return** std_logic_vector;
function "nand" (l, r : std_ulogic_vector) **return** std_ulogic_vector;
function "or"   (l, r : std_logic_vector) **return** std_logic_vector;
function "or"   (l, r : std_ulogic_vector) **return** std_ulogic_vector;
function "nor" (l, r : std_logic_vector) **return** std_logic_vector;
function "nor" (l, r : std_ulogic_vector) **return** std_ulogic_vector;
function "xor" (l, r : std_logic_vector) **return** std_logic_vector;
function "xor" (l, r : std_ulogic_vector) **return** std_ulogic_vector;
function "xnor" (l, r : std_logic_vector) **return** std_logic_vector; -- 1076-
                                                                         -- 1993 only
function "xnor" (l, r : std_ulogic_vector) **return** std_ulogic_vector;

*Note: the **xnor** function is support in 1076-1993 only.*

function "not" (l : std_logic_vector) **return** std_logic_vector;
function "not" (l : std_ulogic_vector) **return** std_ulogic_vector;

Standard Logic Type Conversion Functions

The **std_logic_1164** package includes a variety of type conversion functions to help convert data between 1076 standard data types (**bit** and **bit_vector**) and IEEE 1164 standard logic data types:

```
function To_bit ( s : std_ulogic;   xmap : bit := '0') return bit;
function To_bitvector ( s : std_logic_vector ; xmap : bit := '0') return
bit_vector;
function To_bitvector ( s : std_ulogic_vector; xmap : bit := '0') return
bit_vector;
function To_StdULogic ( b : bit ) return std_ulogic;
function To_StdLogicVector  ( b : bit_vector ) return std_logic_vector;
function To_StdLogicVector  ( s : std_ulogic_vector ) return
std_logic_vector;
function To_StdULogicVector ( b : bit_vector ) return std_ulogic_vector;
function To_StdULogicVector ( s : std_logic_vector  ) return
std_ulogic_vector;
```

Strength Stripping Functions

The strength stripping functions convert the 9-valued types **std_ulogic** and **std_logic** to the 3-, 4-, and 5-valued types (**X01**, **X01Z**, **UX01** and **UX01Z**), converting strength values ('H', 'L', and 'W') to their '0' and '1' equivalents.

```
function To_X01  ( s : std_logic_vector ) return std_logic_vector;
function To_X01  ( s : std_ulogic_vector ) return std_ulogic_vector;
function To_X01  ( s : std_ulogic ) return X01;
function To_X01  ( b : bit_vector ) return std_logic_vector;
function To_X01  ( b : bit_vector ) return std_ulogic_vector;
function To_X01  ( b : bit ) return X01;
function To_X01Z ( s : std_logic_vector  ) return std_logic_vector;
function To_X01Z ( s : std_ulogic_vector ) return std_ulogic_vector;
function To_X01Z ( s : std_ulogic ) return X01Z;
function To_X01Z ( b : bit_vector ) return std_logic_vector;
function To_X01Z ( b : bit_vector ) return std_ulogic_vector;
function To_X01Z ( b : bit ) return X01Z;
function To_UX01  ( s : std_logic_vector  ) return std_logic_vector;
function To_UX01  ( s : std_ulogic_vector ) return std_ulogic_vector;
function To_UX01  ( s : std_ulogic ) return UX01;
```

```
function To_UX01 ( b : bit_vector ) return std_logic_vector;
function To_UX01 ( b : bit_vector ) return std_ulogic_vector;
function To_UX01 ( b : bit ) return UX01;
```

Edge Detection Functions

The edge detection functions **rising_edge()** and **falling_edge()** provide a concise, portable way to describe the behavior of an edge-triggered device such as a flip-flop:

```
function rising_edge  (signal s : std_ulogic) return boolean;
function falling_edge (signal s : std_ulogic) return boolean;
```

Miscellaneous Checking Functions

The following functions can be used to determine if an object or literal is a don't-care, which, for this purpose, is defined as any of the five values **'U'**, **'X'**, **'Z'**, **'W'** or **'-'**:

```
function Is_X ( s : std_ulogic_vector ) return boolean;
function Is_X ( s : std_logic_vector ) return boolean;
function Is_X ( s : std_ulogic ) return boolean;
```

Standard 1076.3 (the Numeric Standard)

IEEE Standard 1076.3 was developed to help synthesis tool users and vendors by providing standard, portable data types and operations for numeric data, and by providing more clearly defined meaning for the nine values of the IEEE 1164 **std_ulogic** and **std_logic** data types.

IEEE Standard 1076.3 provides standard numeric data types.

IEEE Standard 1076.3 defines the package **numeric_std** that allows the use of arithmetic operations on standard logic (**std_logic** and **std_logic_vector**) data types. (The 1076.3 standard also defines arithmetic forms of the **bit** and **bit_vector** data types in a package named **numeric_bit**, but this alternative package is not described here.)

The **numeric_std** package defines the numeric types **signed** and **unsigned** and corresponding arithmetic operations and functions based on the **std_logic** (resolved) data type. The package was designed for use with synthesis tools, and therefore includes additional functions (such as **std_match**) that simplify the use of don't-cares.

Using Numeric Data Types

There are many different applications of the IEEE 1076.3 numeric data types, operators and functions. The following example demonstrates how the type **unsigned** might be used to simplify the description of a counter:

*To make use of the **unsigned** type, you must reference the **numeric_std** package located in the **ieee** library.*

```
-------------------------------------------------------
-- COUNT16: 4-bit counter.
--
library ieee;
use ieee.std_logic_1164.all;
use ieee.numeric_std.all;

entity COUNT16 is
   port (Clk,Rst,Load: in std_logic;
         Data: in std_logic_vector (3 downto 0);
         Count: out std_logic_vector (3 downto 0)
   );
end COUNT16;

architecture COUNT16_A of COUNT16 is
   signal Q: unsigned (3 downto 0);
   constant MAXCOUNT: unsigned (3 downto 0) := "1111";
begin
   process(Rst,Clk)
   begin
     if Rst = '1' then
        Q <= (others => '0');
      elsif rising_edge(Clk) then
        if Load = '1' then
           Q <= UNSIGNED(Data);          -- Type conversion
         elsif Q = MAXCOUNT then
           Q <= (others => '0');
        else
           Q <= Q + 1;
```

```
        end if;
    end if;

    Count <= STD_LOGIC_VECTOR(Q);     -- Type conversion

  end process;

end COUNT16_A;
```

In this example, the type **unsigned** is used within the architecture to represent the counter data. The add operation ('+') is defined for type **unsigned** by the 1076-3 standard (in library **numeric_std**) so the counter can be easily described. Because the **unsigned** and **std_logic_vector** data types share the same element type (**std_logic**), conversion between these types is straightforward, as shown.

Numeric Standard Types: Unsigned and Signed

There are two numeric data types, **unsigned** and **signed**, declared in the **numeric_std** package, as shown below:

```
type unsigned is array (natural range <>) of std_logic;
type signed is array (natural range <>) of std_logic;
```

Unsigned represents unsigned integer data in the form of an array of **std_logic** elements. **Signed** represents signed integer data. In **signed** or **unsigned** arrays, the leftmost bit is treated as the most significant bit. Signed integers are represented in the **signed** array in two's complement form.

Numeric Standard Operators

Arithmetic Operators

```
function "abs" (ARG: signed) return signed;
function "-" (ARG: signed) return signed;
function "+" (L, R: unsigned) return unsigned;
function "+" (L, R: signed) return signed;
function "+" (L: unsigned; R: natural) return unsigned;
function "+" (L: natural; R: unsigned) return unsigned;
```

135

function "+" (L: integer; R: signed) **return** signed;
function "+" (L: signed; R: integer) **return** signed;
function "-" (L, R: unsigned) **return** unsigned;
function "-" (L, R: signed) **return** signed;
function "-" (L: unsigned;R: natural) **return** unsigned;
function "-" (L: natural; R: unsigned) **return** unsigned;
function "-" (L: signed; R: integer) **return** signed;
function "-" (L: integer; R: signed) **return** signed;
function "*" (L, R: unsigned) **return** unsigned;
function "*" (L, R: signed) **return** signed;
function "*" (L: unsigned; R: natural) **return** unsigned;
function "*" (L: natural; R: unsigned) **return** unsigned;
function "*" (L: signed; R: integer) **return** signed;
function "*" (L: integer; R: signed) **return** signed;
function "/" (L, R: unsigned) **return** unsigned;
function "/" (L, R: signed) **return** signed;
function "/" (L: unsigned; R: natural) **return** unsigned;
function "/" (L: natural; R: unsigned) **return** unsigned;
function "/" (L: signed; R: integer) **return** signed;
function "/" (L: integer; R: signed) **return** signed;
function "rem" (L, R: unsigned) **return** unsigned;
function "rem" (L, R: signed) **return** signed;
function "rem" (L: unsigned; R: natural) **return** unsigned;
function "rem" (L: natural; R: unsigned) **return** unsigned;
function "rem" (L: signed; R: integer) **return** signed;
function "rem" (L: integer; R: signed) **return** signed;
function "mod" (L, R: unsigned) **return** unsigned;
function "mod" (L, R: signed) **return** signed;
function "mod" (L: unsigned; R: natural) **return** unsigned;
function "mod" (L: natural; R: unsigned) **return** unsigned;
function "mod" (L: signed; R: integer) **return** signed;
function "mod" (L: integer; R: signed) **return** signed;

Numeric Logical Operators

function "not" (L: unsigned) **return** unsigned;
function "and" (L, R: unsigned) **return** unsigned;
function "or" (L, R: unsigned) **return** unsigned;
function "nand" (L, R: unsigned) **return** unsigned;
function "nor" (L, R: unsigned) **return** unsigned;
function "xor" (L, R: unsigned) **return** unsigned;
function "xnor" (L, R: unsigned) **return** unsigned; -- 1076-1993 only

*Note: the **xnor** operator is not supported in standard 1076-1987.*

function "not" (L: signed) **return** signed;
function "and" (L, R: signed) **return** signed;

function "or" (L, R: signed) **return** signed;
function "nand" (L, R: signed) **return** signed;
function "nor" (L, R: signed) **return** signed;
function "xor" (L, R: signed) **return** signed;
function "xnor" (L, R: signed) **return** signed; -- 1076-1993 only

*Note: the **xnor** operator is not supported in standard 1076-1987.*

Relational Operators

function ">" (L, R: unsigned) **return** boolean;
function ">" (L, R: signed) **return** boolean;
function ">" (L: natural; R: unsigned) **return** boolean;
function ">" (L: integer; R: signed) **return** boolean;
function ">" (L: unsigned; R: natural) **return** boolean;
function ">" (L: signed; R: integer) **return** boolean;
function "<" (L, R: unsigned) **return** boolean;
function "<" (L, R: signed) **return** boolean;
function "<" (L: natural; R: unsigned) **return** boolean;
function "<" (L: integer; R: signed) **return** boolean;
function "<" (L: unsigned; R: natural) **return** boolean;
function "<" (L: signed; R: integer) **return** boolean;
function "<=" (L, R: unsigned) **return** boolean;
function "<=" (L, R: signed) **return** boolean;
function "<=" (L: natural; R: unsigned) **return** boolean;
function "<=" (L: integer; R: signed) **return** boolean;
function "<=" (L: unsigned; R: natural) **return** boolean;
function "<=" (L: signed; R: integer) **return** boolean;
function ">=" (L, R: unsigned) **return** boolean;
function ">=" (L, R: signed) **return** boolean;
function ">=" (L: natural; R: unsigned) **return** boolean;
function ">=" (L: integer; R: signed) **return** boolean;
function ">=" (L: unsigned; R: natural) **return** boolean;
function ">=" (L: signed; R: integer) **return** boolean;
function "=" (L, R: unsigned) **return** boolean;
function "=" (L, R: signed) **return** boolean;
function "=" (L: natural; R: unsigned) **return** boolean;
function "=" (L: integer; R: signed) **return** boolean;
function "=" (L: unsigned; R: natural) **return** boolean;
function "=" (L: signed; R: integer) **return** boolean;
function "/=" (L, R: unsigned) **return** boolean;
function "/=" (L, R: signed) **return** boolean;
function "/=" (L: natural; R: unsigned) **return** boolean;
function "/=" (L: integer; R: signed) **return** boolean;
function "/=" (L: unsigned; R: natural) **return** boolean;
function "/=" (L: signed; R: integer) **return** boolean;

Shift and Rotate Functions

> **function** shift_left (ARG: unsigned; COUNT: natural) **return** unsigned;
> **function** shift_right (ARG: unsigned; COUNT: natural) **return** unsigned;
> **function** shift_left (ARG: signed; COUNT: natural) **return** signed;
> **function** shift_right (ARG: signed; COUNT: natural) **return** signed;
> **function** rotate_left (ARG: unsigned; COUNT: natural) **return** unsigned;
> **function** rotate_right (ARG: unsigned; COUNT: natural) **return** unsigned;
> **function** rotate_left (ARG: signed; COUNT: natural) **return** signed;
> **function** rotate_right (ARG: signed; COUNT: natural) **return** signed;

*Note: the following shift and rotate operators (**sll**, **srl**, **rol** and **ror**) are supported in IEEE 1076-1993 only.*

> **function** "sll" (ARG: unsigned; COUNT: integer) **return** unsigned;
> **function** "sll" (ARG: signed; COUNT: integer) **return** signed;
> **function** "srl" (ARG: unsigned; COUNT: integer) **return** unsigned;
> **function** "srl" (ARG: signed; COUNT: integer) **return** signed;
> **function** "rol" (ARG: unsigned; COUNT: integer) **return** unsigned;
> **function** "rol" (ARG: signed; COUNT: integer) **return** signed;
> **function** "ror" (ARG: unsigned; COUNT: integer) **return** unsigned;
> **function** "ror" (ARG: signed; COUNT: integer) **return** signed;

Numeric Array Resize Functions

The resize functions are used to convert a fixed-sized **signed** or **unsigned** array to a new (larger or smaller) size. If the resulting array is larger than the input array, the result is padded with '0's. In the case of a **signed** array, the sign bit is extended to the least significant bit.

> **function** resize (ARG: signed; NEW_SIZE: natural) **return** signed;
> **function** resize (ARG: unsigned; NEW_SIZE: natural) **return** unsigned;

Numeric Type Conversion Functions

The numeric type conversion functions are used to convert between integer data types and **signed** and **unsigned** data types.

> **function** to_integer (ARG: unsigned) **return** natural;
> **function** to_integer (ARG: signed) **return** integer;

function to_unsigned (ARG, SIZE: natural) **return** unsigned;
function to_signed (ARG: integer; SIZE: natural) **return** signed;

Matching Functions

The matching functions (**std_match**) are used to determine if two values of type **std_logic** are logically equivalent, taking into consideration the semantic values of the 'X' (uninitialized) and '-' (don't-care) literal values. The table of Figure 4-3 (derived from the **match_table** constant declaration found in the **numeric_std** package) defines the matching of all possible combinations of the **std_logic** enumerated values.

	'U'	'X'	'0'	'1'	'Z'	'W'	'L'	'H'	'-'
'U'	F	F	F	F	F	F	F	F	T
'X'	F	F	F	F	F	F	F	F	T
'0'	F	F	T	F	F	F	T	F	T
'1'	F	F	F	T	F	F	F	T	T
'Z'	F	F	F	F	F	F	F	F	T
'W'	F	F	F	F	F	F	F	F	T
'L'	F	F	T	F	F	F	T	F	T
'H'	F	F	F	T	F	F	F	T	T
'-'	T	T	T	T	T	T	T	T	T

Figure 4-3. The *std_match* function compares two values and determines if they are logically equivalent.

function std_match (L, R: STD_ULOGIC) **return** boolean;
function std_match (L, R: unsigned) **return** boolean;
function std_match (L, R: signed) **return** boolean;
function std_match (L, R: std_logic_vector) **return** boolean;
function std_match (L, R: STD_ULOGIC_vector) **return** boolean;

Translation Functions

The numeric translation functions convert the nine **std_logic** values to numeric binary values ('0' or '1') for use in **signed** and **unsigned** arithmetic operations. These translation functions convert the values of 'L' and 'H' to '0' and '1', respectively. Any other values ('U', 'X', 'Z', '-', or 'W') result in a warning error (assertion) being generated.

function to_01 (S: unsigned; XMAP: std_logic := '0') **return** unsigned;
function to_01 (S: signed; XMAP: std_logic := '0') **return** signed;

139

We're Cookin' Now...

It's starting to come together, isn't it? We've got standard data types, operators and functions to work with, and we've seen some simple examples of how these data types are used. Now let's explore in more detail some of the issues we touched on in Chapter 2, *A First Look At VHDL*. We'll start by ...

5. Understanding Concurrent Statements

In Chapter 2, *A First Look At VHDL*, we covered the three primary design description styles (behavior, dataflow and structure) supported by VHDL. We also discussed the difference between writing VHDL statements that are concurrent (written within the body of an architecture declaration) and those that are sequential (written within a VHDL process statement, function or procedure). We looked at examples of all three description styles and also demonstrated, by example, that there is not a direct relationship between (a) concurrent and sequential statements in VHDL, and (b) combinational and sequential (registered) logic.

VHDL is a concurrent language, meaning that it is possible to describe parallel operations using the language.

In this chapter, we will examine more closely the concept of concurrency as it is implemented in VHDL and VHDL simulators. We will also explore some of the concurrent language features of VHDL in more detail and learn how combinational and registered logic can be described using these features. In addition, we will look briefly at how timing delays are annotated to concurrent assignments in VHDL, so you will have a better understanding of how simulation models are constructed. (This book does not cover simulation modeling in detail, as there are many fine books already written on this subject. Besides, relatively few of today's VHDL users are actually creating new simulation models themselves.)

The Concurrent Area

*Concurrent state-ments are normally entered between the **begin** and **end** statements of the architecture.*

In VHDL, there is only one place where you will normally enter concurrent statements. This place, the *concurrent area*, is found between the **begin** and **end** statements of an architecture declaration. The following VHDL diagram shows where the concurrent area of a VHDL architecture is located:

```
architecture arch1 of my_circuit is
    signal Reset, DivClk: std_logic;
    constant MaxCount: std_logic_vector(15 downto 0) := "10001111";
    component count port (Clk, Rst: in std_logic;
                          Q: out std_logic_vector(15 downto 0));
    begin

        Reset <= '1' when Qout = MaxCount else '0';

        CNT1: count port map(GClk, Reset, DivClk);

        Control: process(DivClk)
        begin
          . . .
        end process;
          . . .
    end arch1;
```

Concurrent Area

All statements within the concurrent area are considered to be parallel in their execution and of equal priority and importance. Processes (described in more detail in Chapter 6, *Understanding Sequential Statements*) also obey this rule, executing in parallel with other assignments and processes appearing in the concurrent area.

There is no order dependency to statements in the concurrent area, so the following architecture declaration:

```
architecture arch1 of my_circuit is
    signal A, B, C: std_logic_vector(7 downto 0);
    constant Init: std_logic_vector(7 downto 0) := "01010101";
begin
    A <= B and C;
    B <= Init when Select = '1' else C;
    C <= A and B;
end arch1;
```

is exactly equivalent to:

```
architecture arch2 of my_circuit is
    signal A, B, C: std_logic_vector(7 downto 0);
    constant Init: std_logic_vector(7 downto 0) := "01010101";
begin
    C <= A and B;
    A <= B and C;
    B <= Init when Select = '1' else C;
end arch2;
```

There is no order dependency between concurrent statements.

The easiest way to understand this concept of concurrency is to think of concurrent VHDL statements as a kind of *netlist*, in which the various assignments being made are nothing more than connections between different types of *objects*. If you think in terms of a schematic, you might mentally create a picture for the preceding description that looks like the schematic of Figure 5-1.

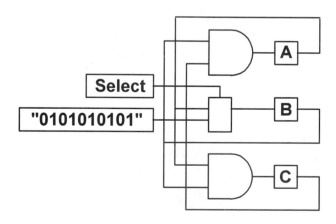

Figure 5-1: *It is convenient to think of concurrent statements in terms of a schematic netlist.*

This is not a particularly useful circuit, but it illustrates the point: if you think of the signals, constants, components, literals—and even processes—available in concurrent VHDL statements as distinct objects (such as you might find on a schematic or block diagram), and think of operations (such as **and**, **not**, and **when-else**) and assignments as logic gates and wiring specifications, respectively, then you will have no trouble understanding how VHDL's concurrent statements can be mapped to actual digital logic.

Concurrent Signal Assignments

The most common and simple concurrent statements are concurrent signal assignments...

The most common and simple concurrent statements you will write in VHDL are *concurrent signal assignments*. Concurrent signal assignments such as those shown in the previous section specify the logical relationships between different signals in a digital system.

If you have used PLD-oriented design languages (such as PALASM, ABEL, CUPL or Altera's AHDL), then concurrent signal assignments will be quite familiar to you. Just like the Boolean equations that you write using a PLD language, concurrent signal assignments in VHDL describe logic that is inherently parallel.

Because all signal assignments in your design description are concurrent (including those described within processes, as we will see in the next chapter), there is no relevance to the order in which the assignments are made within the concurrent area of the architecture.

In most cases, you will use concurrent signal assignments to describe either combinational logic (using logic expressions of arbitrary complexity), or you will use them to describe the connections between lower-level components. In some cases (though not typically for designs that will be synthesized) you will use concurrent signal assignments to describe registered logic as well.

The following example includes two simple concurrent signal assignments that represent NAND and NOR operations:

```
architecture arch3 of nand_circuit is
    signal A, B: std_logic;
    signal Y1, Y2: std_logic;
begin
    Y1 <= not (A and B);
    Y2 <= not (A or B);
end arch3;
```

In this example (as in the example presented earlier), there is no significance to the order in which the two assignments have been made. Also, keep in mind that the two signals being assigned (**Y1** and **Y2**) could just as easily have been ports of the entity rather than signals declared in the architecture. In all cases, signals declared locally (within an architecture, for example) can be used in exactly the same ways as can ports of the corresponding entity. The only difference between ports and locally-declared signals is that ports have a direction, or *mode* (**in**, **out** or **inout**), limiting whether they can have values assigned to them (in the case of **in**), or whether they can be read as inputs (in the case of **out**). If a port is declared as mode **out**, its value cannot be read. It can only be assigned a value. A port of mode **in** is the opposite; it can be read, but it cannot be assigned a value. A port of mode **inout** has both capabilities.

Signal assignments can also include delay specifications, as described later in this chapter.

Conditional Signal Assignment

Conditional signal assignments allow you to describe a sequence of related conditions and corresponding signal assignments.

A conditional signal assignment is a special form of signal assignment, similar to the if-then-else statements found in software programming languages, that allows you to describe a sequence of related conditions under which one or more signals are assigned values. The following example (a simple multiplexer) demonstrates the basic form of a conditional assignment:

```
entity my_mux is
   port (Sel: in std_logic_vector (0 to 1);
            A, B, C, D: in std_logic_vector (0 to 3);
            Y: out std_logic_vector (0 to 3));
end my_mux;

architecture mux1 of my_mux is
begin

   Y <= A when Sel  = "00" else
        B when Sel = "01" else
        C when Sel = "10" else
        D when others;

end mux1;
```

*The **others** clause provides a default assignment feature.*

A conditional signal assignment consists of an assignment to one output (or a collection of outputs, such as an array of any type) and a series of conditional **when** statements, as shown. To ensure that all conditions are covered, you can use a terminating **when others** clause, as was done for the multiplexer description above.

Note:

*It is very important that all conditions in a conditional assignment are covered, as unwanted latches can be easily generated from synthesis for those conditions that are not covered. In the preceding multiplexer example, you might be tempted to replace the clause **D when others** with **D when Sel = "11"** (to improve readability). This would not be correct, however, because the data type being used in the design (**std_logic_vector**) has nine possible values for each bit. This means that there are actually 81 possible unique values that the input **Sel** could have at any given time, rather than four.*

The conditional signal assignment also provides a concise method of describing a list of conditions that have some priority. In the case of the multiplexer just described, there is no priority required or specified, since the four conditions (the possible values of the 2-bit input **Sel**) are all mutually exclusive. In some design descriptions, however, the priority implied by a series of **when-else** statements can cause some confusion (and additional logic being generated). For this reason, you might want to use a selected signal assignment (described in the next section) as an alternative.

Selected Signal Assignment

A selected signal assignment is similar to a conditional signal assignment (described in the previous section) but differs in that the input conditions specified have no implied priority. The following is an example of a selected signal assignment:

```
entity my_mux is
    port (Sel: in std_logic_vector (0 to 1);
          A, B, C, D: in std_logic_vector (0 to 3);
          Y: out std_logic_vector (0 to 3));
end my_mux;

architecture mux1 of my_mux is
begin

    with Sel select
      Y <= A when "00",
           B when "01",
           C when "10",
           D when others;

end mux1;
```

A selected signal assignment does not imply a priority for the selection conditions.

In this simple multiplexer example, the selected signal assignment has exactly the same function as the conditional signal assignment presented earlier. This is not always the case, however, and you should carefully evaluate which type of assignment is most appropriate for a given application.

Conditional vs. Selected Assignment

How to choose between a conditional assignment and a selected assignment? Consider this: a conditional assignment always enforces a priority on the conditions. For example, the conditional expression:

Conditional signal assignments can result in unwanted additional logic being generated from synthesis.

```
Q1 <= "01" when A = '1' else
      "10" when B = '1' else
      "11" when C = '1' else
      "00";
```

is identical to the selected assignment:

```
with std_logic_vector'(A,B,C) select
  Q2 <= "01" when "100",
        "01" when "101",
        "01" when "110",
        "01" when "111",
        "10" when "010",
        "10" when "011",
        "11" when "001",
        "00" when others;
```

Notice that input **A** takes priority. In the conditional assignment, that priority is implied by the ordering of the expressions. In the selected assignment, you must specify all possible conditions, so there can be no priority implied.

Why is this important for synthesis? Consider a circuit in which we know in advance that only one of the three inputs (**A**, **B**, or **C**) could ever be active at the same time. Or perhaps we don't care what the output of our circuit is under the condition where more than one input is active. In such cases, we can reduce the amount of logic required for our design by eliminating the priority implied by the conditional expression. We could instead write our description as:

A selected signal assignment will never result in additional priority logic being generated.

```
with std_logic_vector'(A,B,C) select
  Q2 <= "01" when "100",
        "10" when "010",
        "11" when "001",
        "00" when others;
```

This version of the description will, in all likelihood, require less logic to implement than the earlier version. This kind of optimization can save dramatic amounts of logic in larger designs.

In summary, while a conditional assignment may be more natural to write, a selected signal assignment may be preferable to avoid introducing additional, unwanted logic in your circuit.

Other notes

- You must include all possible conditions in a selected assignment. If not all conditions are easily specified, you can use the **others** clause as shown above to provide a default assignment.

- The selection expressions may include ranges and multiple values. For example, you could specify ranges for a **bit_vector** selection expression as follows:

```
with Address select
    CS <= SRAM when 0x"0000" to 0x"7FFF",
          PORT when 0x"8000" to 0x"81FF",
          UART when 0x"8200" to 0x"83FF",
          PROM when others;
```

- VHDL `93 adds the following feature to the selected signal assignment: You can use the keyword **unaffected** to specify that the output does not change under one or more conditions. For example, a multiplexer with two selector inputs could be described as:

```
with Sel select
    Y <= A when "00",
         B when "01",
         C when "10",
         unaffected when others;
```

Synthesis Note:

*The preceding multiplexer description may result in a latch being generated from synthesis. This is because the synthesized circuit will have to maintain the value of the output **Y** when the value of input **Sel** is "11".*

149

Procedure Calls

Procedures may be called concurrently within an architecture. When procedures are called concurrently, they must appear as independent statements within the concurrent area of the architecture.

Concurrent procedure calls are very much like processes...

You can think of procedures in the same way you think of processes within an architecture: as independent sequential programs that execute whenever there is a change (an event) on any of their inputs. The advantage of a procedure over a process is that the body of the procedure (its sequential statements) can be kept elsewhere (in a package, for example) and used repeatedly throughout the design.

In the following example, the procedure **dff** is called within the concurrent area of the architecture:

```
architecture shift2 of shift is
    signal D,Qreg: std_logic_vector(0 to 7);
begin

    D <= Data when (Load = '1') else
            Qreg(1 to 7) & Qreg(0);

    dff(Rst, Clk, D, Qreg);

    Q <= Qreg;

end shift2;
```

Generate Statements

Generate statements can be used to create repetitive logic.

Generate statements are provided as a convenient way to create multiple instances of concurrent statements, most typically component instantiation statements. There are two basic varieties of **generate** statements.

The for-generate statement

The following example shows how you might use a **for-generate** statement to create four instances of a lower-level component (in this case a RAM block):

```
architecture generate_example of my_entity is
    component RAM16X1
        port(A0, A1, A2, A3, WE, D: in std_logic;
            O: out std_logic);
    end component;
begin
    . . .
    RAMGEN: for i in 0 to 3 generate
        RAM: RAM16X1 port map ( . . . );
    end generate;
    . . .
end generate_example;
```

When this **generate** statement is evaluated, the VHDL compiler will generate four unique instances of component **RAM16X1**. Each instance will have a unique name that is based on the instance label provided (in this case **RAM**) and the index value.

For-generate statements can be nested, so it is possible to generate multi-dimensional arrays of component instances or other concurrent statements.

The if-generate statement

The **if-generate** statement is most useful when you need to conditionally generate a concurrent statement. A typical example of this occurs when you are generating a series of repetitive statements or components and need to supply different parameters, or generate different components, at the beginning or end of the series. The following example shows how a combination of a **for-generate** statement and two **if-generate** statements can be used to describe a 10-bit parity generator constructed of cascaded exclusive-OR gates:

```
library ieee;
use ieee.std_logic_1164.all;
```

```
entity parity10 is
    port(D: in std_logic_vector(0 to 9);
         ODD: out std_logic);
    constant width: integer := 10;
end parity10;

library gates;
use gates.all;

architecture structure of parity10 is
    component xor2
        port(A,B: in std_logic;
             Y: out std_logic);
    end component;
    signal p: std_logic_vector(0 to width - 2);
begin
    G: for I in 0 to (width - 2) generate
        G0: if I = 0 generate
            X0: xor2 port map(A => D(0), B => D(1), Y => p(0));
        end generate G0;
        G1: if I > 0 and I < (width - 2) generate
            X0: xor2 port map(A => p(i-1), B => D(i+1), Y => p(i));
        end generate G1;
        G2: if I = (width - 2) generate
            X0: xor2 port map(A => p(i-1), B => D(i+1), Y => ODD);
        end generate G2;
    end generate G;
end structure;
```

*Using a combination of **for generate** and **if generate** statements, it is possible to create large, regular structures.*

Concurrent Processes

*Taken as a whole, a **process** statement is one large concurrent statement within an architecure.*

Process statements contain sequential statements but are themselves concurrent statements within an architecture. In most VHDL design descriptions, there are multiple processes that execute concurrently during simulation and describe hardware that is inherently concurrent in its operation.

In the following example, two processes are used to describe a background clock (process CLOCK) and a sequence of stimulus inputs in a test bench:

```vhdl
architecture Stim1 of TEST_COUNT4EN is

component COUNT4EN
    port ( CLK,RESET,EN : in  std_logic;
        COUNT : out std_logic_vector(3 downto 0)
    );
end component;

constant CLK_CYCLE : Time := 20 ns;

signal CLK,INIT_RESET,EN : std_logic;
signal COUNT_OUT : std_logic_vector(3 downto 0);

begin
    U0: COUNT4EN port map ( CLK=>CLK,RESET=>INIT_RESET,
                EN=>EN, COUNT=>COUNT_OUT);
    process begin
        CLK <= '1';
        wait for CLK_CYCLE/2;
        CLK <= '0';
        wait for CLK_CYCLE/2;
    end process;

    process begin
        INIT_RESET <= '0'; EN <= '1';
        wait for  CLK_CYCLE/3;
        INIT_RESET <= '1';
        wait for  CLK_CYCLE;
        INIT_RESET <= '0';
        wait for  CLK_CYCLE*10;
        EN <= '0';
        wait for  CLK_CYCLE*3;
        EN <= '1';

        wait;
    end process;
end Stim1;
```

Multiple processes within an architecure operate independent of each other, and in parallel.

The interrelationships between multiple processes in a design description can be complex. They are discussed in Chapter 6, *Understanding Sequential Statements*. For the purpose of understanding concurrency, however, you must never assume that any process you write will be executed in simulation prior to any other process. This means that you cannot count on signals or shared variables being updated between two processes.

Component Instantiations

Component instantiations are used to create unique references to lower-level components.

Component instantiations are statements that reference lower-level components in your design, in essence creating unique copies (or *instances*) of those components. A component instantiation statement is a concurrent statement, so there is no significance to the order in which components are referenced. You must, however, declare any components that you reference in either the declarative area of the architecture (before the **begin** statement) or in an external package that is visible to the architecture.

The following example demonstrates how component instantiations can be written. In this example, there are two lower-level components (**half_adder** and **full_adder**) that are referenced in component instantiations to create a total of four component instances. When simulated or synthesized, the four component instances (**A0**, **A1**, **A2** and **A3**) will be processed as four independent circuit elements. In this example, we have declared the two lower-level components **half_adder** and **full_adder** right in the architecture. To make your design descriptions more concise, you may choose to place component declarations in separate packages instead.

Components must be declared before they are referenced in a component instantiation.

```
library ieee;
 use ieee.std_logic_1164.all;
entity adder4 is
   port(A,B: in std_logic_vector(3 downto 0);
        S: out std_logic_vector(3 downto 0);
        Cout: out std_logic);
end adder4;

architecture structure of adder4 is
   component half_adder
      port (A, B: in std_logic; Sum, Carry: out std_logic);
   end component;
   component full_adder
```

```
        port (A, B, Cin: in std_logic; Sum, Carry: out std_logic);
    end component;
    signal C: std_logic_vector(0 to 2);
begin

    A0: half_adder port map(A(0), B(0),      S(0), C(0));
    A1: full_adder port map(A(1), B(1), C(0), S(1), C(1));
    A2: full_adder port map(A(2), B(2), C(1), S(2), C(2));
    A3: full_adder port map(A(3), B(3), C(2), S(3), Cout);

end structure;
```

Port mapping

Port mapping can be described using either positional or named association.

The mapping of ports in a component can be described in one of two ways. The simplest method (and the method used in the preceding example) is called *positional association*. Positional association simply maps signals in the architecture (the *actuals*) to corresponding ports in the lower-level entity declaration (the *formals*) by their position in the port list. When using positional association, you must provide exactly the same number and types of ports as are declared for the lower-level entity.

Positional association is quick and easy to use, and it is tempting to use this method almost exclusively. However, there are potential problems with positional association. The most troublesome problem is the lack of error checking. It is quite easy, for example, to inadvertently reverse the order of two ports in the list. The result is a circuit that may compile with no errors, but fail to simulate properly. After the first few times you accidentally swap the reset and clock lines to one of your lower-level components, you may decide that it is worth the extra typing to provide a more complete specification of your port mappings. The method you will use in this case is called *named association*.

Named association is an alternate form of port mapping that includes both the actual and formal port names in the port map of a component instantiation. (Named association can also be used in other places, such as in the parameter lists for generics and subprograms.)

We could modify the previous 4-bit adder example to use named association as follows:

```
architecture structure of adder4 is
    component half_adder
        port (A, B: in std_logic; Sum, Carry: out std_logic);
    end component;
    component full_adder
        port (A, B, Cin: in std_logic; Sum, Carry: out std_logic);
    end component;
    signal C: std_logic_vector(0 to 2);
begin

    A0: half_adder port map(A => A(0), B => B(0), Sum => S(0),
        Carry => C(0));
    A1: full_adder port map(A => A(1), B => B(1), Cin => C(0),
        Sum => S(1), Carry => C(1));
    A2: full_adder port map(A => A(2), B => B(2), Cin => C(1),
        Sum => S(2), Carry => C(2));
    A3: full_adder port map(A => A(3), B => B(3), Cin => C(2),
        Sum => S(3), Carry => Cout);

end structure;
```

Named association helps to eliminate confusing and hard-to-debug connection errors.

When you specify port mappings using named association, lower-level names (the *formal* ports of the component) are written on the left side of the **=>** operator, while the top-level names (the *actuals*) are written on the right.

The benefits of named association go beyond simple error checking. Because named association removes the requirement for any particular order of the ports, you can enter them in whatever order you want. You can even leave one or more ports unconnected if you have provided default values in the lower-level component specification.

Because named association is so much more flexible (and less error prone) than positional association, we strongly recommend that you get in the habit of typing in the few extra characters required to use named association.

Generic Mapping

If the lower-level entity being referenced includes generics (described in more detail in Chapter 8, *Partitioning Your Design*), you can specify a generic map in addition to the port map to pass actual generic parameters to the lower-level entity:

```
architecture timing of adder4 is
    component half_adder
        port (A, B: in std_logic; Sum, Carry: out std_logic);
    end component;
    component full_adder
        port (A, B, Cin: in std_logic; Sum, Carry: out std_logic);
    end component;
    signal C: std_logic_vector(0 to 2);
begin

    A0: half_adder
        generic map(tRise => 1 ns, tFall => 1 ns);
        port map(A => A(0), B => B(0), Sum => S(0), Carry => C(0));
    A1: full_adder
        generic map(tRise => 1 ns, tFall => 1 ns);
        port map(A => A(1), B => B(1), Cin => C(0), Sum => S(1),
                Carry => C(1));
    A2: full_adder
        generic map(tRise => 1 ns, tFall => 1 ns);
        port map(A => A(2), B => B(2), Cin => C(1), Sum => S(2),
                Carry => C(2));
    A3: full_adder
        generic map(tRise => 1 ns, tFall => 1 ns);
        port map(A => A(3), B => B(3), Cin => C(2), Sum => S(3),
                Carry => Cout);

end timing;
```

Generics can be used to pass additional information into a component.

Just as with port maps, generic maps can be written using either positional or named association.

Note:

The rules of VHDL allow you to mix positional and named association in the same port, generic or parameter list. Doing so has little or no benefit, however, and it may confuse other potential users of your design description.

Delay Specifications

VHDL allows signal assignments to include delay specifications, in the form of an **after** clause. The **after** clause allows you to model the behavior of gate and wire delays in a circuit. This is very useful if you are developing simulation models or if you want to include estimated delays in your synthesizable design description. The following are two examples of delay specifications associated with signal assignments:

VHDL provides two basic methods for attaching delay information to a signal assignment.

Y1 <= **not** (A **and** B) **after 7 ns**;

Y2 <= **not** (A **and** B) transport **after 7 ns**;

These two assignments demonstrate the two fundamental types of delay specifications available in VHDL: *inertial* and *transport*.

Inertial delay is intended to model the delay through a gate, in which there is some minimum pulse length that must be maintained before an event is propogated. The timing diagram of Figure 5-2 shows how the output **Y1** (above) would respond to two different length pulses on inputs **A** and **B**:

Figure 5-2: Inertial delay models the behavior of a gate, in which pulses of short duration are not propogated.

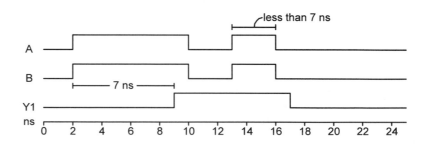

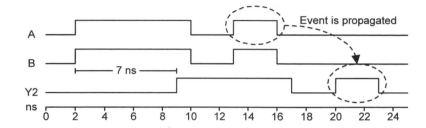

Figure 5-3: *Transport delay models the behavior of a wire, in which all pulses (events) are propogated.*

Transport delay, on the other hand, models the delay on a wire, so pulses of any width are propogated, as shown in Figure 5-3.

Estimated delay information can be added to a synthesizable design description, or actual delays can be back-annotated later in the design process.

For design descriptions intended for synthesis, you will probably not bother to use delay specifications such as these. A circuit produced as a result of synthesis is unlikely to have timing characteristics that can be accurately predicted (or specified) up front. In fact, all synthesis tools in use as of this writing ignore the **after** clause completely. (If you have a general idea of the timing characteristics of your synthesis target—be it an FPGA chip or a high-complexity ASIC—you can use delay specifications to improve the accuracy of your initial simulation. Just be aware that anything you annotate prior to synthesis will be little more than a guess.)

When you are writing test benches, you will also probably not use **after** clauses to specify timing of input events. Instead, you will likely rely on a series of **wait** statements entered within a process to accurately specify your test stimulus.

VHDL `93 Note:

*The IEEE 1076-1993 standard adds an additional feature called a **reject** time. For inertial delays (the default delay type if transport is not speci-fied), a minimum inertial pulse time can be specified as follows:*

Y1 <= **reject** 3 ns **not** (A **and** B) **after** 7 ns;

159

*In this example, any event greater than 3 ns in width will be propagated to the output. In the absence of a specified reject time, the specified delay time (in this case 7 ns) is used as the default **reject** time.*

Signal Drivers

VHDL includes an elaborate set of rules and language features to resolve situations in which the same signal is driven to multiple values simultaneously. These situations can be caused unintentionally (by an incomplete or incorrect design specification), or they may represent a desired circuit condition, such as a three-state driver connected to a bus, or they may represent a simple output enable used in a loadable bi-directional register.

Every signal assignment results in a signal driver being created.

To handle such situations, VHDL introduces the concept of a *signal driver*. A signal driver is a conceptual circuit that is created for every signal assignment in your circuit. By default, this conceptual circuit provides a comparison function to ensure that only one driver is active at any given time. The following architecture demonstrates a circuit description that does not meet this requirement:

```
architecture arch4 of nand_circuit is
    signal Sel, A, B: std_logic;
    signal Y: std_logic;
begin
    Y <= not (A and B) and Sel;
    Y <= not (A or B) and not Sel;
end arch4;
```

The intent of this circuit is to provide a single output (**Y**) that functions either as a NAND gate or as a NOR gate based on the value of **Sel**. Unfortunately, each of the two assignments results in a driver being created, resulting in a multiple-driver situation.

The solution to this, of course, is to completely specify the output **Y** using only one signal assignment, as in the following:

```
architecture arch4 of nand_circuit is
   signal Sel, A, B: std_logic;
   signal Y,Y1,Y2: std_logic;
begin
   Y1 <= not (A and B);
   Y2 <= not (A or B);
   Y <= Y1 and Sel or Y2 and not Sel;
end arch4;
```

In this example, two intermediate signals have been introduced (**Y1** and **Y2**) and the output **Y** has been more completely described as a function of these two values. Another method might be to simply combine the three assignments into a larger combinational expression (**not (A and B) and Sel or not (A or B) and not Sel**) or to use a more concise statement such as a conditional assignment:

```
architecture arch5 of nand_circuit is
   signal Sel, A, B: std_logic;
   signal Y,Y1,Y2: std_logic;
begin
   Y <= not (A and B) when Sel = '1' else
        not (A or B);
end arch5;
```

VHDL's signal driver rules can help you to detect and correct logic errors in your design.

Of course, these simple examples only show how you might resolve multiple driver situations that have been inadvertently created. You will find that VHDL's signal driver rules can actually help to detect and correct errors in your design that might otherwise go unnoticed. For situations that are intentional, however, how can you get around the rules? The answer is a feature of VHDL called a *resolution function*. A resolution function is a special type of function that you (or someone else, such as the IEEE committee that defined the resolved type **std_logic**) can write to resolve multiple-driver situations for a specific type. For example, the resolution function for a four-value data type consisting of the values '1', '0', 'X' (unknown) and 'Z' (high impedance) might have a resolution function that specifies:

1) that simultaneous values of '1' and '0' appearing on a signal's drivers will result in an 'X' value,

2) that both 'Z' and 'X' can be over-ridden by values of '1' or '0', and

3) that 'Z' is over-ridden by 'X'.

Chapter 3, *Exploring Objects and Data Types*, includes an example of a resolution function. That chapter also describes the concepts of signal drivers, resolved types and resolution functions in more detail. For most design descriptions and test benches, however, you will not need to use resolved types such as these. (In many synthesis tools, resolution functions are not supported anyway. They serve only to let the compiler know whether multiple drivers are allowed for an output.)

Now, On To...

As you can see, concurrent statements are a powerful aspect of VHDL, allowing you to describe an inherently parallel system. But in circuit design (as in life), sometimes you have to put things in their proper order. We can do that in VHDL by...

6. Understanding Sequential Statements

In the previous chapter, we examined the features of VHDL that allow concurrent events to be described. Concurrent VHDL allows you to describe the operation of an inherently parallel system such as a combinational logic network or other digital logic circuit. However, concurrent VHDL does not allow you to clearly specify, at a higher level, what that system is supposed to do *over time*.

Sequential statements allow you to describe your circuit as a sequence of related events...

Sequential VHDL statements, on the other hand, allow you to describe the operation, or *behavior*, of your circuit as a sequence of related events. Such descriptions are natural for order-dependent circuits such as state machines and for complex combinational logic that involves some priority of operations. The use of sequential statements to describe combinational logic implies that our use of the term *sequential* in VHDL is somewhat different from the term as it is often used to describe digital logic. Specifically, sequential statements written in VHDL do not necessarily represent sequential digital logic circuits. As we will see, it is possible (and quite common) to write sequential VHDL statements, using processes and subprograms, to describe what is essentially combinational logic.

In this chapter we will look at examples of both registered logic and combinational logic described using sequential statements. We will also examine the various types of sequential statements available in VHDL. Our primary focus will be on those styles of sequential VHDL that are most appropriate for synthesizable design descriptions and for test benches. We will also touch on issues related to delay specifications and the order in which processes are analyzed—issues closely related to sequential VHDL.

Sequential statements are found within processes, functions and procedures.

Sequential statements are found within processes, functions, and procedures. Sequential statements differ from concurrent statements in that they have *order dependency*. This order dependency may or may not imply a sequential circuit (one involving memory elements).

The Process Statement

VHDL's **process** statement is the primary way you will enter sequential statements. A process statement, including all declarations and sequential statements within it, is actually considered to be a single concurrent statement within a VHDL architecture. This means that you can write as many processes and other concurrent statements as are necessary to describe your design, without worrying about the order in which the simulator will process each concurrent statement.

Anatomy of a Process

The general form of a process statement is:

*A **process** statement typically includes a sensitivity list, local declarations, and one or more sequential statements.*

```
process_name: process (sensitivity_list)
    declarations
begin
    sequential_statements
end process;
```

The easiest way to think of a VHDL process is to relate it to event-driven software—like a program that executes (in simulation) any time there is an event on one of its inputs (as specified in the sensitivity list). A process describes the sequential execution of statements that are dependent on one or more events having occurred. A flip-flop is a perfect example of such a situation. It remains idle, not changing state, until there is a significant event (either a rising edge on the clock input or an asynchronous reset event) that causes it to operate and potentially change its state.

Signals that are assigned a value within a process will hold their values between executions of that process.

Although there is a definite order of operations within a process (from top to bottom), you can think of a process as executing in zero time. This means that a process can be used to describe circuits functionally, without regard to their actual timing, and multiple processes can be "executed" in parallel with little or no concern for which processes complete their operations first.

A process can be thought of as a single concurrent statement written within a VHDL architecture, extending from the **process** keyword (or from the optional process name that precedes it) to the terminating **end process** keyword pair and semicolon.

The process name (*process_name*) appearing before the **process** keyword is optional and can be used to: (1) identify specific processes that are executing during simulation, and (2) more clearly distinguish elements such as local variables that may have common names in different processes.

Immediately following the **process** statement is an optional list of signals enclosed by parentheses. This list of signals, called the sensitivity list, specifies the conditions under which the process is to begin executing. When a sensitivity list is associated with a process, any change in the value of any input in the list will result in immediate execution of the process.

*A process must have either a sensitivity list or a **wait** statement, but not both.*

In the absence of a sensitivity list, the process will execute continuously, but must be provided with at least one **wait** statement to cause the process to suspend periodically. Examples of processes that are written with and without sensitivity lists are presented in the subsections below.

The order in which statements are written in a process is significant. You can think of a process as a kind of software program that is executed sequentially, from top to bottom, each time it is invoked during simulation. Consider, for example, the following process describing the operation of a counter:

```
process(Clk)
begin
   if Clk = '1' and Clk'event then
      if Load = '1' then
         Q <= Data_in;
      else
         Q <= Q + 1;
      end if;
   end if;
end process;
```

When this process is executed, the statements appearing between the **begin** and **end process** statements are executed in sequence. In this example, the first statement is an **if** test that will determine if there was a rising edge on the **Clk** clock input. A second, nested **if** test determines if the counter should be loaded with **Data_in** or incremented, depending on the value of the **Load** input.

When is a process invoked? That depends on the type of process it is. There are two fundamental types of processes that you can write: those that have sensitivity lists and those that do not.

Processes With Sensitivity Lists

A process with a sensitivity list is executed during simulation whenever an event occurs on any of the signals in the sensitivity list. An event is defined as any change in value of a

166

signal, such as when a signal of type **Boolean** changes from **True** to **False**, or when the value of an **integer** type signal is incremented or otherwise modified.

Processes that include sensitivity lists are most often used to describe the behavior of circuits that respond to external stimuli. These circuits, which may be either combinational, sequential (registered), or a combination of the two, are normally connected with other sub-circuits or interfaces, via signals, to form a larger system. In a typical circuit application, such a process will include in its sensitivity list all inputs that have asynchronous behavior. These inputs may include clocks, reset signals, or inputs to blocks of combinational logic.

The following is an example of a process that includes a sensitivity list. This process describes the operation of a clocked shift register with an asynchronous reset; note the use of the `event signal attribute to determine which of the two signals (**Clk** and **Rst**) had an event:

A process with a sensitivity list executes only when one of the signals in the sensitivity list changes its value.

```
process(Rst, Clk)
begin
  if Rst = '1' then
    Q <= "00000000";
  elsif Clk = '1' and Clk'event then
    if Load = '1' then
      Q <= Data_in;
    else
      Q <= Q(1 to 7) & Q(0);
    end if;
  end if;
end process;
```

During simulation, whenever there is an event on either **Rst** or **Clk**, this process statement will execute from the **begin** statement to the **end process** statement pair. If the **Rst** input is '1' (regardless of whether the event that triggered the process execution was **Rst** or **Clk**), then the output **Q** is set to a reset value of "00000000". If the value of **Rst** is not '1', then the **Clk** input is checked to determine if it has a value of '1' and had an event. This checking for both a value and an event is a common (and synthesizable) way of detecting transitions, or edges, on signals such as clocks.

After all of the statements in the process have been analyzed and executed, the process is suspended until a new event occurs on one of the process' sensitivity list entries.

For design descriptions intended for input to synthesis software, you should follow the above example and write process statements that include sensitivity lists, as this is the most widely used synthesis convention for registers.

Processes Without Sensitivity Lists

Processes without sensitivity lists execute continuously, suspending only when a wait statement is encountered.

A process that does not include a sensitivity list executes somewhat differently than a process with a sensitivity list. Rather than executing from the **begin** statement at the top of the process to the **end process** statement, a process with no sensitivity list executes from the beginning of the process to the first occurrence of a **wait** statement, then suspends until the condition specified in the **wait** statement is satisfied. If the process only includes a single **wait** statement, the process reactivates when the condition is satisfied and continues to the **end process** statement, then begins executing again from the beginning. If there are multiple **wait** statements in the process, the process executes only until the next **wait** statement is encountered.

The following example demonstrates how this works, using a simplified manchester encoder as an example:

```
process
begin
   wait until Clk = '1' and Clk'event;
   M_out <= data_in;
   wait until Clk = '1' and Clk'event;
   M_out <= not data_in;
end process;
```

This process will suspend its execution at two points. The first **wait until** statement suspends the process until there is a rising edge on the clock (a transition to a value of '1'). When this rising edge condition has been met, the process continues execution by assigning the value of **data_in** to **M_out**. Next, the second **wait until** statement suspends the process until another rising edge has been detected on **Clk.** When this condition has been met, the process continues and assigns the inverted value of **data_in** to **M_out**. The process does not suspend at the **end process** statement, but instead loops back to the beginning and immediately starts processing over again.

The use of multiple **wait** statements within a process makes it possible to describe very complex multiple-clock circuits and systems. Unfortunately, such design descriptions usually fall outside of the scope of today's synthesis tools. Rather than use multiple **wait** statements to describe such logic, you will probably use **wait** statements only when describing test stimulus, as discussed later in this chapter.

Using Processes for Combinational Logic

In the previous chapter, we saw how concurrent signal assignments can be used to create combinational logic. When you write a sequence of concurrent signal assignments, each statement that you write is independent of all other statements and results in a unique combinational function (unless a guarded block or some other special feature is used to imply memory).

Processes can be used to describe combinational logic as well as registered logic.

If you wish, you can use sequential VHDL statements (in the form of a process or subprogram) to create combinational logic as well. Sequential VHDL statements can actually be more clear and concise for many types of combinational functions, as they allow the priority of operations to be clearly expressed within a combinational logic function.

169

The following is an example of a simple combinational logic function (a 4-into-1 multiplexer) described using a process:

```
entity simple_mux is
    port (Sel: in bit_vector (0 to 1);
            A, B, C, D: in bit;
            Y: out bit);
end simple_mux;

architecture behavior of simple_mux is
begin
    process(Sel, A, B, C, D)
    begin
        if Sel = "00" then
            Y <= A;
        elsif Sel = "01" then
            Y <= B;
        elsif Sel = "10" then
            Y <= C;
        elsif Sel = "11" then
            Y <= D;
        end if;
    end process;
end simple_mux;
```

A process describes combinational logic when all inputs are listed in the sensitivity list and there are no undefined input conditions.

This simple process describes combinational logic because it conforms to the following rules:

1. The sensitivity list of the process includes all signals that are being read (i.e., used as inputs) within the process.

2. Assignment statements written for the process outputs (in this case only output **Y**) cover all possible combinations of the process inputs (in this case **Sel, A, B, C** and **D**).

These two rules dictate whether the signal assignment logic generated from a process is strictly combinational or will require some form of memory element (such as a flip-flop or latch).

For processes that include variable declarations, there is an additional rule that comes into play:

170

3. All variables used in the process must have a value assigned to them before they are read (i.e., used as inputs).

An example of when an apparently combinational logic description actually describes registered logic is demonstrated by the modified (6-into-1) multiplexer description shown below:

```
entity simple_mux is
   port (Sel: in bit_vector (0 to 2);
         A, B, C, D, E, F: in bit;
         Y: out bit);
end simple_mux;

architecture behavior of simple_mux is
begin
   process(Sel, A, B, C, D, E, F)
   begin
     if Sel = "000" then
        Y <= A;
     elsif Sel = "001" then
        Y <= B;
     elsif Sel = "010" then
        Y <= C;
     elsif Sel = "011" then
        Y <= D;
     elsif Sel = "100" then
        Y <= E;
     elsif Sel = "101" then
        Y <= F;
     end if;
   end process;
end simple_mux;
```

Registered logic can be accidentally created when one or more input conditions are left undefined.

This modified version of the multiplexer has only six of the eight possible values for **Sel** described in the **if-then-elsif** statement chain. What happens when **Sel** has a value of **"110"** or **"111"**? Unlike many simpler hardware description languages (most notably languages such as ABEL or CUPL that are intended for programmable logic use), the default behavior in VHDL is to hold the values of unspecified signals. For output **Y** to hold its value when **Sel** has a value of **"110"** or

171

"111", a memory element (such as a latch) will be required. The result is that the circuit as described is no longer a simple combinational logic function.

Understanding what types of design descriptions will result in combinational logic and what types will result in latches and flip-flops is very important when writing VHDL for synthesis. For more information, see Appendix A, *Getting The Most Out Of Synthesis*.

Using Processes for Registered Logic

Process statements are ideal for describing registered circuits.

Perhaps the most common use of VHDL processes is to describe the behavior of circuits that have memory and must save their state over time. The sequential nature of VHDL processes (and subprograms) make them ideal for the description of such circuits.

In the previous section we listed three rules that must be obeyed in order to ensure that the circuitry described by a process is combinational. If your goal is to create registered logic (using either flip-flop or latch elements), then you will describe your design using one or more of the following methods:

1. Write a process that does not include all of its inputs in the sensitivity list.

2. Use incompletely specified **if-then-elsif** logic to imply that one or more signals must hold their values under certain conditions.

3. Use one or more variables in such a way that they must hold a value between iterations of the process. (For example, specify a variable as an input to an assignment before that variable has been assigned a value itself.)

To ensure the highest level of compatibility with synthesis tools, you should use a combination of methods 1 and 2. The following example demonstrates how registered logic (the shift register presented in Chapter 2, *A First Look At VHDL*) can be described using a process:

```
-------------------------------------------------------
-- Eight-bit  shifter
--
library ieee;
use ieee.std_logic_1164.all;
entity rotate is
    port( Clk, Rst, Load: in std_logic;
            Data: in std_logic_vector(0 to 7);
            Q: out std_logic_vector(0 to 7));
end rotate;

architecture rotate1 of rotate is
    signal Qreg: std_logic_vector(0 to 7);
begin
    reg: process(Rst,Clk)
    begin
        if Rst = '1' then   -- Async reset
            Qreg <= "00000000";
        elsif (Clk = '1' and Clk'event) then
            if (Load = '1') then
                Qreg <= Data;
            else
                Qreg <= Qreg(1 to 7) & Qreg(0);
            end if;
        end if;
    end process;
    Q <= Qreg;
end rotate1;
```

*An incomplete **if-then** statement implies that signal **Qreg** must hold its value under certain conditions.*

In this example, the incomplete **if-then** statement implies that signal **Qreg** will hold its value when the two conditions (a reset or clock event) are false.

For a detailed explanation of this example, see Chapter 2, *A First Look At VHDL.*

Using Processes for State Machines

State machines are common circuits that are easy to describe using processes.

State machines are a common form of sequential logic circuits that are used for generating or detecting sequences of events. To describe a synthesizable state machine in VHDL, you should follow a well-established coding convention that makes use of enumerated types and processes. The following example demonstrates how to write a synthesizable state machine description using this coding convention.

The circuit, a video frame grabber controller, was first described (in the form of an ABEL language description) in *Practical Design Using Programmable Logic* by David Pellerin and Michael Holley (Prentice Hall, 1990).

The circuit described is a simple freeze-frame unit that grabs and holds a single frame of NTSC color video image. This design description includes the frame detection and capture logic. The complete circuit requires an 8-bit D-A/A-D converter and a 256K X 8 static RAM.

The design description makes use of a number of independent processes. The first process (which has been given the name of **ADDRCTR**), describes a large counter corresponding to the frame address counter in the circuit. This counter description makes use of the IEEE Standard 1076.3 numeric data type **unsigned**.

The second process, **SYNCCTR**, also describes a counter using the **unsigned** data type. This counter is used to detect the vertical blanking interval, which indicates the start of one frame of video.

The third and fourth processes (**STREG** and **STTRANS**) describe the operation of the video frame grabber controller logic, using the most common (and most easily synthesized) form for state machines. First, an enumerated type called states is declared that consists of the values **StateLive**, **StateWait**, **StateSample**, and **StateDisplay**. Two intermediate signals (**current_state** and **next_state**) are then introduced to represent the current state and calculated next state of the machine. In the processes that follow, signal **current_state** represents a set of state registers, while **next_state** represents a combinational logic function.

The diagram of Figure 6-1 illustrates the operation of the video frame grabber controller:

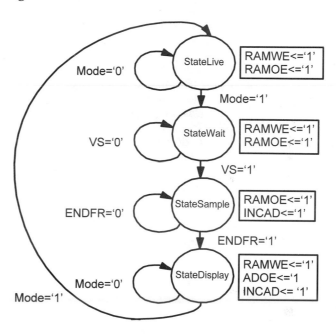

Figure 6-1: The video frame grabber controller is represented as a 4-state state machine.

Process **STREG** describes the operation of the state registers, and simply loads the value of the calculated next state (signal **next_state**) into the state registers (**current_state**) whenever there is a synchronous clock event. This process also includes asynchronous reset logic that will set the machine to its initial state (**StateLive**) when the **Rst** input is asserted.

The actual transition logic for the state machine is described in process **STTRANS**. In this process, a **case** statement is used to decode the current state of the machine (as represented by

175

signal **current_state**) and define the transitions between states. This is an example where sequential VHDL statements are used to describe non-sequential (combinational) logic.

```
--------------------------------------------------------
-- A Video Frame Grabber.
--
library ieee;
use  ieee.std_logic_1164.all;
use  ieee.numeric_std.all;

entity video is
   port (Reset, Clk: in std_logic;
       Mode: in std_logic;
       Data: in std_logic_vector(7 downto 0);
       TestLoad: in std_logic;
       Addr: out std_logic_vector(17 downto 0);
       RAMWE: out std_logic;
       RAMOE: out std_logic;
       ADOE: out std_logic);
end video;

architecture control1 of video is
   constant FRAMESIZE: integer := 253243;
   constant TESTADDR: integer := 253000;

   signal ENDFR: std_logic;
   signal INCAD: std_logic;
   signal VS: std_logic;
   signal Sync: unsigned (6 downto 0);
begin

   -- Address counter. This counter increments until we reach the end of
   -- the frame (address 253243), or until the input INCAD goes low.

   ADDRCTR: process(Clk)
      variable cnt: unsigned (17 downto 0);
   begin
      if rising_edge(Clk) then
         if TestLoad = '1' then
            cnt := to_unsigned(TESTADDR,18);
            ENDFR <= '0';
         else
            if INCAD = '0' or cnt = FRAMESIZE then
               cnt := to_unsigned(0,18);
```

```vhdl
      else
         cnt := cnt + to_unsigned(1,18);
      end if;
      if cnt = FRAMESIZE then
         ENDFR <= '1';
      else
         ENDFR <= '0';
      end if;
    end if;
  end if;
  Addr <= std_logic_vector(cnt);
end process;


-- Vertical sync detector. Here we look for 128 bits of zero, which
-- indicates the vertical sync blanking interval.
SYNCCTR: process(Reset,Clk)
begin
  if Reset = '1' then
    Sync <= to_unsigned(0,7);
  elsIf rising_edge(Clk) then
    if Data /= "00000000" or Sync = 127 then
      Sync <= to_unsigned(0,7);
    else
      Sync <= Sync + to_unsigned(1,7);
    end if;
  end if;
end process;


VS <= '1' when Sync = 127 else '0';


STATEMACHINE: block
  type states is (StateLive,StateWait,StateSample,StateDisplay);
  signal current_state, next_state: states;
begin
  -- State register process:
  STREG: process(Reset,Clk)
  begin
    if Reset = '1' then
      current_state <= StateLive;
    elsif rising_edge(Clk) then
      current_state <= next_state;
    end if;
  end process;

  -- State transitions:
  STTRANS: process(current_state,Mode,VS,ENDFR)
```

```
begin
  case current_state is
    when StateLive =>    -- Display live video on the output
      RAMWE <= '1';
      RAMOE <= '1';
      ADOE <= '0';
      INCAD <= '0';
      if Mode = '1' then
        next_state <= StateWait;
      else
        next_state <= StateLive
      end if;
    when StateWait =>    -- Wait for vertical sync
      RAMWE <= '1';
      RAMOE <= '1';
      ADOE <= '0';
      INCAD <= '0';
      if VS = '1' then
        next_state <= StateSample;
      else
        next_state <= StateWait
      end if;
    when StateSample => -- Sample one frame of video
      RAMWE <= '0';
      RAMOE <= '1';
      ADOE <= '0';
      INCAD <= '1';
      if ENDFR = '1' then
        next_state <= StateDisplay;
      else
        next_state <= StateSample
      end if;
    when StateDisplay => -- Display the stored frame
      RAMWE <= '1';
      RAMOE <= '0';
      ADOE <= '1';
      INCAD <= '1';
      if Mode = '1' then
        next_state <= StateLive;
      else
        next_state <= StateDisplay
      end if;
  end case;
end process;
end block;
end control1;
```

Specifying State Machine Encodings

The state encoding that you select for your state machine can have a dramatic impact on the design size.

We have described the preceding video frame grabber in an implementation-independent fashion, with the assumption that whatever synthesis tool we use to process this design will come up with an optimal solution, in terms of the state encodings selected. For small designs such as this, or when you are not tightly constrained for space, it is probably fine to let the synthesis tool encode your states for you. In many cases, however, you will have to roll up your sleeves and work on improving the synthesis results yourself, by creating your own optimizal state encodings. Determining an optimal encoding for a large state machine can be a long and tedious process, the methods for which are beyond the scope of this book. It is important to understand the various coding styles for manually-encoded machines, however, to get the most out of synthesis.

Using Constants for State Encodings

The easiest way to specify an explicit encoding for a state machine is to replace the declaration and use of an enumerated type with a series of constant declarations. For the video frame grabber, for example, we could replace the declarations:

```
type states is (StateLive,StateWait,StateSample,StateDisplay);
signal current_state, next_state: states;
```

with:

Constants can be used to precisely specify the encoding of a state machine.

```
type states is std_logic_vector(1 downto 0);
constant StateLive: states := "00";
constant StateWait: states := "01";
constant StateSample: states := "11";
constant StateDisplay: states := "10";
signal current_state, next_state: states;
```

Using these declarations will result in the precise encodings that we have specified in the synthesized circuit. There is one additional modification we will have to make to our frame grabber state machine if we specify the states using declarations based on **std_logic_vector**, however. Because the base type of **std_logic_vector** (**std_logic**) has nine unique values, the four constants that we have declared (**StateLive**, **StateWait**, **StateSample** and **StateDisplay**) do not represent all possible values for the state type. For this reason, we will have to add an **others** clause to the **case** statement describing the transitions of our machine, as in:

```
when others =>
    null;
```

Using the Enum_encoding Synthesis Attribute

An alternate method of specifying state machine encodings is provided in some synthesis tools, inluding Synopsys, Exemplar, and Metamor. This method makes use of a non-standard (but widely supported) attribute called **enum_encoding**. The following modified declarations (again, using the video frame grabber state machine as an example) uses the **enum_encoding** attribute to specify the same state encoding used in the previous example:

*The **enum_encoding** attribute is commonly used to specify state encodings.*

```
type states is (StateLive,StateWait,StateSample,StateDisplay);
attribute enum_encoding of states: type is "00 01 11 10";
signal current_state, next_state: states;
```

The **enum_encoding** attribute used in this example has been defined elsewhere (most probably in a special library package provided by the synthesis vendor) as a string:

```
attribute enum_encoding: string;
```

This attribute is recognized by the synthesis tool, which encodes the generated state machine circuitry accordingly. During simulation, the **enum_encoding** attribute is ignored, and we will instead see the enumerated values displayed.

Specifying a One-hot Encoding

A one-hot encoding can dramatically reduce logic requirements.

One common technique for optimizing state machine logic is to use what is called a *one-hot* encoding, in which there is one register dedicated to each state in the machine. One-hot machines require more register resources than more typical, *maximally-encoded* machines, but can result in tremendous savings in the combinational logic required for next-state and output decoding. This trade-off can be particularly effective in device technologies that have an abundance of built-in registers, but that suffer from limited (or relatively slow) routing resources.

When you first try to use a one-hot approach to state encoding, it is tempting to describe the machine using the same methods that you might have used for your other state machines. The following declarations represent an attempt to encode our video frame grabber state machine one-hot using constant declarations:

Simply specifying a single-bit-true encoding will not result in optimal logic being produced.

```
type states is std_logic_vector(3 downto 0);
constant StateLive: states := "0001";
constant StateWait: states := "0010";
constant StateSample: states := "0100";
constant StateDisplay: states := "1000";
signal current_state, next_state: states;
```

At first glance this looks correct; each state is represented by a single bit being asserted, and when simulated and synthesized, the machine will indeed transition to the appropriate encoded state for each transition described in the **case** statement shown earlier. In terms of the logic required for state decoding, however, we have not achieved a genuine one-hot machine. This is because the **case** statement we have written describing the state transitions implicitly refers to all four state registers when decoding the current state of the machine. A true, optimal one-hot machine only requires that one register be observed to determine if the machine is in a given state.

To generate the correct logic, optimized as a one-hot encoded machine, we have to modify the description somewhat, so that only one state register is examined for each possible transition. The easiest way to do this is to replace the **case** statement with a series of **if** statements, as follows:.

A more optimal solution is to reference only the bit representing a given state.

```
-- State transitions for one-hot encoding:
STTRANS: process(current_state,Mode,VS,ENDFR)
begin
   if current_state(0) = '1' then    -- StateLive
        RAMWE <= '1';
        RAMOE <= '1';
        ADOE <= '0';
        INCAD <= '0';
        if Mode = '1' then
           next_state <= StateWait;
        else
           next_state <= StateLive
        end if;
   end if;
   if current_state(1) = '1' then    -- StateWait
        RAMWE <= '1';
        RAMOE <= '1';
        ADOE <= '0';
        INCAD <= '0';
        if VS = '1' then
           next_state <= StateSample;
        else
           next_state <= StateWait
        end if;
   end if;
   if current_state(2) = '1' then    -- StateSample
        RAMWE <= '0';
        RAMOE <= '1';
        ADOE <= '0';
        INCAD <= '1';
        if ENDFR = '1' then
           next_state <= StateDisplay;
        else
           next_state <= StateSample
        end if;
   end if;
   if current_state(3) = '1' then    -- StateDisplay
        RAMWE <= '1';
        RAMOE <= '0';
```

```
          ADOE <= '1';
          INCAD <= '1';
          if Mode = '1' then
             next_state <= StateLive;
          else
             next_state <= StateDisplay
          end if;
       end if;
    end process;
```

We could, of course, make this description more readable by introducing constants for the index values for each state register.

Using Processes for Test Stimulus

Processes are frequently used to describe test stimulus.

In addition to their use for describing combinational and registered circuits to be synthesized or modeled for simulation, VHDL processes are also important for describing the test environment in the form of sequential application of stimulus and (if desired) checking of resulting circuit outputs.

A process that is intended for testing (as part of a test bench) will normally have no sensitivity list. Instead, it will have a series of **wait** statements that provide time for the unit under test to stabilize between the assignment of test inputs. Because a process intended for use as a test bench does not describe hardware to be synthesized, you are free to use any legal features and style of VHDL without regard to the limitations of synthesis.

The following is a simplistic test bench example written with a single process statement. This process statement might be used to apply a sequence of input values to a lower-level circuit and check the state of that circuit's outputs at various points in time.

```
    -- A simple process to apply various stimulus over time...
    process
       constant PERIOD: time := 40 ns;
```

```
begin
  Rst <= '1';
  A <= "00000000";
  B <= "00000000";
  wait for PERIOD;
  CheckState(Q, "00000000");
  Rst <= '0';
  A <= "10101010";
  B <= "01010101";
  wait for PERIOD * 4;
    CheckState(Q, "11111111");
  A <= "11111010";
  B <= "01011111";
  wait for PERIOD * 2;
    CheckState(Q, "00110101");
  wait;
end process;
```

In this example, the process executes just once before suspending indefinitely (as indicated by the final **wait** statement). The stimulus is described by a sequence of assignments to signals **A** and **B**, and by calls to a procedure (defined elsewhere) named **CheckState**. **Wait** statements are used to describe a delay between each test sequence.

More comprehensive examples of using processes for test stimulus can be found in Chapter 9, *Writing Test Benches*.

Sequential Statements in Subprograms

All statements entered within a procedure or function are sequential.

We've seen examples of how sequential statements are written in a process statement. The process statement is relatively easy to understand if you think of it as a small software program that executes independent of other processes and concurrent statements during simulation.

Functions and procedures (which are collectively called subprograms) are very similar to processes in that they contain sequential statements executed as independent 'programs' during simulation. The parameters you pass into a subprogram are analogous to the sensitivity list of a process;

whenever there is an event on any object (signal or variable) being passed as an argument to a subprogram, that subprogram is executed and its outputs (whether they are output parameters, in the case of a procedure, or the return value, in the case of a function) are recalculated.

The following example includes a procedure declared within the architecture. The procedure counts the number of ones and zeroes there are in a **std_logic_vector** input (of arbitrary width) and returns the count values as output parameters. The procedure is used to build two result strings containing the appropriate number of ones and zeroes, left justified and padded with 'X' values. (For example, an input with the values "1010001001" would result in the values "1111XXXXXX" and "000000XXXX".)

```
entity proc is
    port (Clk: in std_logic;
          Rst: in std_logic;
          InVector: in std_logic_vector(0 to 9);
          OutOnes: out std_logic_vector(0 to 9);
          OutZeroes: out std_logic_vector(0 to 9));
end proc;

architecture behavior of proc is

    procedure CountBits(InVector: in std_logic_vector;
                        ones,zeroes: out natural) is
        variable cnt1: natural := 0;
        variable cnt0: natural := 0;
    begin
        for I in 0 to InVector'right loop
            case InVector(I) is
                when '1' => cnt1 := cnt1 + 1;
                when '0' => cnt0 := cnt0 + 1;
                when others => null;
            end case;
        end loop;
        ones := cnt1;
        zeroes := cnt0;
    end CountBits;

    signal Tmp1, Tmp0: std_logic_vector(0 to 9);
begin
```

Procedures and functions can be declared within an architecture.

185

```
process(Rst, Clk)
begin
  if Rst = '1' then
    OutOnes <= (others => '0');
    OutZeroes <= (others => '0');
  elsif rising_edge(Clk) then
    OutOnes <= Tmp1;
    OutZeroes <= Tmp0;
  end if;
end process;

process(InVector)
  variable ones, zeroes: natural;
begin
  Countbits(InVector,ones,zeroes);
  Tmp0 <= (others => 'X');
  Tmp1 <= (others => 'X');
  for I in 0 to ones - 1 loop
    Tmp1(I) <= '1';
  end loop;
  for I in 0 to zeroes - 1 loop
    Tmp0(I) <= '0';
  end loop;
end process;

end behavior;
```

This example shows that a procedure containing sequential statements can be invoked from within a process—or even from within another procedure. The calling process simply suspends until the procedure has completed execution.

Note:

*This example is theoretically synthesizable, but the fact that the procedure has been written without regard to the width of the inputs will probably make it impossible to process by synthesis tools. If this design were to be synthesized, the variables **cnt1** and **cnt0** would have to be given range constraints.*

Signal and Variable Assignments

It is important to understand the scheduling rules for signals assigned within processes and subprograms.

One important aspect of VHDL you should clearly understand is the relationship between sequential statements (in a process or subprogram) and the scheduling of signal and variable assignments. Signals within processes have fundamentally different behavior from variables. Variables are assigned new values immediately, while signal assignments are scheduled and do not occur until the current process (or subprogram) has been suspended. When you describe complex logic using sequential assignments, you must carefully consider which type of object (signal or variable) is appropriate for that part of your design.

Let's look at an example where signal assignments would be appropriate. In this example, an 8-bit serial cyclic-redundancy-check (CRC) generator, signals are required because we are attempting to construct a chain of registers. Each register in the chain is clocked from a common source, and data moves from one register to the next only when there is an event on **Clk**. We might describe the data as being "scheduled."

```
-------------------------------------------------------------------------
-- 8-bit Serial CRC Generator.
--
library ieee;
use ieee.std_logic_1164.all;

entity crc8s is
   port (Clk,Set, Din: in std_logic;
       CRC_Sum: out std_logic_vector(15 downto 0));
end crc8s;

architecture behavior of crc8s is
   signal X: std_logic_vector(15 downto 0);
begin
   process(Clk,Set)
   begin
     if Set = '1' then
       X <= "1111111111111111";
     elsif rising_edge(Clk) then
       X(0)  <= Din xor X(15);
       X(1)  <= X(0);
```

```
            X(2)  <= X(1);
            X(3)  <= X(2);
            X(4)  <= X(3);
            X(5)  <= X(4) xor Din xor X(15);
            X(6)  <= X(5);
            X(7)  <= X(6);
            X(8)  <= X(7);
            X(9)  <= X(8);
            X(10) <= X(9);
            X(11) <= X(10);
            X(12) <= X(11) xor Din xor X(15);
            X(13) <= X(12);
            X(14) <= X(13);
            X(15) <= X(14);
         end if;
       end process;
     CRC_Sum <= X;
 end behavior;
```

Because the data moving from register to register is scheduled, this example would not work if **X** was described using a variable instead of a signal. If a variable was substituted for **X**, the assignments for each stage of the CRC generation would be immediate and thus would not describe a chain of registers.

Note also that the assignment of **X** to **CRC_Sum** must be placed outside the process. If we were to write the assignment to **CRC_Sum** within the process, as in:

```
       . . .
          X(14) <= X(13);
          X(15) <= X(14);
       end if;

     CRC_Sum <= X;

   end process;

end behavior;
```

the result would not be what we intend. This is because the assignment of **CRC_Sum** will be subject to the execution and signal assignment rules of a process. In this case, the assignment of a final value to **X** will be delayed until the process suspends, and **CRC_Sum** will not be updated until the next time the process executes. (As it turns out, the next time the process executes may well be on the falling edge of the clock, meaning that **CRC_Sum** would be delayed by half a clock cycle.)

Signals and Variables: Which To Use?

Deciding whether to use signals or variables requires careful consideration...

Deciding whether to use a signal or a variable for a given object can require some thought and careful consideration of the requirements of your design. The wrong choice can lead to a significant amount of wasted time during simulation and debugging.

Assuming your design is a traditional synchronous system, there is a way to simplify your decision somewhat. If you step back mentally and consider your design in terms of registers and combinational logic (taking a dataflow view of the design), you can think of the difference between signals and variables as follows. If your design is composed of semi-independent registers that are arranged as shown in Figure 6-2, then you should use signal assignments and write something that looks like this:

```
architecture version1 of circuit is
    signal Q1, Q2: std_logic;
begin
    process (Clk)
    begin
        if rising_edge(Clk) then
            Q1 <= A and B;
            Q2 <= Q1 and C;
        end if;
    end process;
end version1;
```

However, if you are trying to describe a more parallel circuit in which the registers are not fed back in a chain (Figure 6-3), or if you don't want registers generated for every assignment in the process, then you will want to use variable assignments, and write something like this:

```
architecture version1 of circuit is
   signal Q1, Q2: std_logic;
begin
   process (Clk)
      variable tmp1, tmp2: std_logic;
   begin
      if rising_edge(Clk) then
         tmp1 := A and B;
         tmp2 := Q1 and C;
      end if;
      Q1 <= tmp1;
      Q2 <= tmp2;
   end process;
end version1;
```

Figure 6-2: *Multiple signal assignments within a process can be used to describe a chain of registers.*

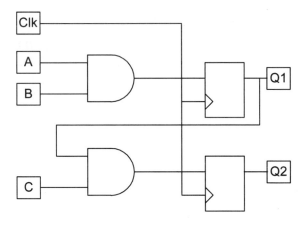

Figure 6-3: *The same design described using variables might not result in the desired logic.*

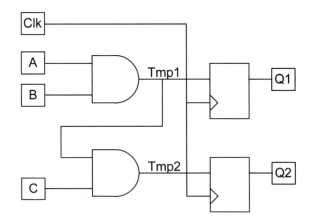

In reality, describing complex synchronous circuits in VHDL usually requires some combination of signals and variables. The preceding examples should help you to understand the differences between these two types of objects.

If-Then-Else Statements

VHDL includes a variety of *control statements* that can be used to describe combinational functions, indicate priorities of operations, and specify other high-level behavior.

*The **if-then-else** construct is the most common form of control statement in VHDL...*

The **if-then-else** construct is the most common form of control statement in VHDL. The general form of the **if-then-else** construct is:

```
if first_condition then
    statements
elsif second_condition then
    statements
else
    statements
end if;
```

The conditions specified in an **if-then-else** construct must evaluate to a Boolean type. This means that the following example is incorrect:

```
procedure Mux(signal A, B, S: in std_logic; signal O: out std_logic) is
begin
    if S then          -- Error:  S is not Boolean!
        O <= B;
    else
        O <= A;
    end if;
end Mux;
```

Instead, this example must be modified so that the **if** statement condition evaluates to a Boolean expression:

```
if S = '1' then        -- Now it will work...
    O <= B;
else
    O <= A;
```

191

```
        end if;
    end Mux;
```

The statement parts of an **if-then-else** construct can contain any sequential VHDL statements, including other **if-then-else** statement constructs. This means that you can nest multiple levels of **if-then-else** statements, in the following form:

```
if outer_condition then
    statements
else
    if inner_condition then
        statements
    end if;
end if;
```

Case Statements

Case statements are a type of control statement that can be used as alternatives to **if-then-else** constructs. **Case** statements have the following general form:

```
case control_expression is
    when test_expression1 =>
        statements
    when test_expression2 =>
        statements
    when others =>
        statements
end case;
```

Case statements must include all possible conditions of the control expression...

The test expressions of a **case** statement must be mutually exclusive, meaning that no two test expressions are allowed to be true at the same time. **Case** statements must also include all possible conditions of the control expression. (The **others** expression can be used to guarantee that all conditions are covered.)

The primary difference between descriptions written using **case** statements from those written using **if-then-else** statements is that **if-then-else** statements imply a priority of conditions, while a **case** statement does not imply any priority.

(This is similar to the difference between the conditional and selected assignments described in Chapter 5, *Understanding Concurrent Statements*.)

Loops

Loop statements are another catagory of control structures that allow you to specify repeating sequences of behavior in a circuit. There are three primary types of loops in VHDL: **for** loops, **while** loops, and infinite loops.

For Loop

The **for** loop is a sequential statement that allows you to specify a fixed number of iterations in a behavioral design description. The following architecture demonstrates how a simple 8-bit parity generator can be described using a **for** loop:

```
library ieee;
use ieee.std_logic_1164.all;

entity parity10 is
    port(D: in std_logic_vector(0 to 9);
        ODD: out std_logic);
    constant WIDTH: integer := 10;
end parity10;

architecture behavior of parity10 is
begin

    process(D)
        variable otmp: Boolean;
    begin
        otmp := false;
        for i in 0 to D'length - 1 loop
            if D(i) = '1' then
                otmp := not otmp;
            end if;
        end loop;
        if otmp then
            ODD <= '1';
```

```
      else
          ODD <= '0';
      end if;
   end process;

end behavior;
```

The **for** loop includes an automatic declaration for the index (**i** in this example). You do not need to separately declare the index variable.

The index variable and values specified for the loop do not have to be numeric types and values. In fact, the index range specification does not even have to be represented by a range. Instead, it can be represented by a type or sub-type indicator. The following example shows how an enumerated type can be used in a loop statement:

Non-numeric types can be used for loop index variables...

```
architecture looper2 of my_entity is
    type stateval is Init, Clear, Send, Receive, Error;    -- States of a machine
begin
    . . .
    process(a)
    begin
      for state in stateval loop
        case state is
          when Init =>
              ...
          when Clear =>
              ...
          when Send =>
              ...
          when Receive =>
              ...
          when Error =>
              ...
        end case;
      end loop;
    end process;
    . . .
end looper2;
```

For loops can be given an optional name, as shown in the following example:

```
loop1: for state in stateval loop
    if current_state = state  then
        valid_state <= true;
    end if;
end loop loop1;
```

The loop name can be used to help distinguish between the loop index variable and other similarly-named objects, and to specify which of the multiple nested loops is to be terminated (see *Loop Termination* below). Otherwise, the loop name serves no purpose.

While Loop

A **while** loop is another form of sequential loop statement that specifies the conditions under which the loop should continue, rather than specifying a discrete number of iterations. The general form of the **while** loop is shown below:

```
architecture while_loop of my_entity is
begin
    . . .
    process(. . .)
    begin
        . . .
        loop_name: while (condition) loop
            -- repeated statements go here
        end loop loop_name;
        . . .
    end process;
    . . .
end while_loop;
```

Like the **for** loop, a **while** loop can only be entered and used in sequential VHDL statements (i.e., in a process, function or procedure). The loop name is optional.

The following example uses a **while** loop to describe a constantly running clock that might be used in a test bench. The loop causes the clock signal to toggle with each loop iteration, and the loop condition will cause the loop to terminate if either of two flags (**error_flag** or **done**) are asserted.

```
process
begin
  while error_flag /= '1' and done /= '1' loop
    Clock <= not Clock;
    wait for CLK_PERIOD/2;
  end loop;
end process;
```

Note:

*Although **while** loops are quite useful in test benches and simulation models, you may have trouble if you attempt to synthesize them. Synthesis tools may be unable to generate a hardware representation for a **while** loop, particularly if the loop expression depends on non-static elements such as signals and variables. Because support for **while** loops varies widely among synthesis tools, we recommend that you not use them in synthesizable design descriptions.*

Infinite Loop

An infinite loop is a loop statement that does not include a **for** or **while** iteration keyword (or *iteration scheme*). An infinite loop will usually include an **exit** condition, as shown in the template below:

```
architecture inifinite_loop of my_entity is
begin
  . . .
  process(. . .)
    . . .
    loop_name: loop
      . . .
      exit when (condition);
    end loop loop_name;
  end process;
  . . .
end infinite_loop;
```

An infinite loop using a **wait** statement is shown in the example below. This example exhibits exactly the same behavior as the **while** loop shown previously:

```
process
begin
  loop
    Clock <= not Clock;
```

```
    wait for CLK_PERIOD/2;
    if done = '1' or error_flag = '1' then
        exit;
    end if;
  end loop;
end process;
```

As with a **while** loop, an infinite loop probably has no equivalent in hardware and is therefore not synthesizable.

Loop Termination

*An **exit** statement can be used to terminate a loop.*

There are many possible reasons for wanting to jump out of a loop before its normal terminating condition has been reached. The three types of loops previously described all have the ability to be terminated prematurely. Loop termination is performed through the use of an **exit** statement. When an **exit** statement is encountered, its condition is tested and, if the condition is true, the simulator skips the remaining statements in the loop and all remaining loop iterations, and continues execution at the statement immediately following the **end loop** statement.

The following example demonstrates how loop termination can be used to halt a sequence of test vectors that are being executed when an error is detected:

```
for i in 0 to VectorCount loop
    ApplyVector(InputVec(i), ResultVec);
    exit when CheckOutput(OutputVec(i), ResultVec) = FatalError;
end loop;
```

The exit condition is optional; an **exit** statement without an exit condition will unconditionally terminate when the **exit** statement is encountered. The following example shows an unconditional exit termination specified in combination with an **if-then** statement to achieve the same results as in the previous example:

```
for i in 0 to VectorCount loop
    ApplyVector(InputVec(i), ResultVec);
```

```
        if CheckOutput(OutputVec(i), ResultVec) = FatalError then
            exit;
    end loop;
```

When multiple loops are nested, the **exit** statement will terminate only the innermost loop. If you need to terminate a loop that is not the innermost loop, you can make use of loop labels to specify which loop is being terminated. The following example shows how loop labels are specified in **exit** statements:

```
LOOP1: while (StatusFlag = STATUS_OK) loop
    GenerateSequence(InputVec,OutputVec,VectorCount,Seed);
    LOOP2: for i in 0 to VectorCount loop
        ApplyVector(InputVec(i), ResultVec);
        ErrStatus := CheckOutput(OutputVec(i), ResultVec) = TestError;
        if ErrStatus = ERR_COMPARE then
            ReportError();
            exit LOOP2;
        elsif ErrStatus = ERR_FATAL then
            ReportFatal();
            exit LOOP1;
        end if;
    end loop LOOP2;
end loop LOOP1;
```

Next Step: The Joys Of Recycling

Now that we've seen how to use **if**s, **then**s, **loop**s, and other sequential statements, let's capitalize on all of these VHDL capabilities. One great way to do this—and improve your productivity at the same time—is by...

7. Creating Modular Designs

Modular (or structured) programming is a technique that you can use to enhance your own design productivity, as well as that of your design team. A modular design approach allows commonly-used segments of VHDL code to be re-used. It also enhances design readability.

VHDL includes many features that can help you create modular designs. In this chapter we will look at features that allow you to quickly and easily create reusable segments of your design, based on methods similar to those used in software programming languages.

Functions and Procedures

Functions and procedures are useful features for creating modular designs.

Functions and procedures in VHDL, which are collectively known as *subprograms*, are directly analogous to functions and procedures in a high-level software programming language such as C or Pascal. A procedure is a subprogram that has an argument list consisting of inputs and outputs, and no return value. A function is a subprogram that has only inputs in its argument list, and has a return value.

Subprograms are useful for isolating commonly-used segments of VHDL source code. They can either be defined locally (within an architecture, for example), or they can be placed in a package and used globally throughout the design description or project.

Statements within a subprogram are sequential...

Statements within a subprogram are sequential (like a process), regardless of where the subprogram is invoked. Subprograms can be invoked from within the concurrent area of an architecture or from within a sequential process or higher-level subprogram. They can also be invoked from within other subprograms.

Subprograms are very much like processes in VHDL. In fact, any statement that you can enter in a VHDL process can also be entered in a function or procedure, with the exception of a **wait** statement (since a subprogram executes once each time it is invoked and cannot be suspended while it is executing). It is therefore useful to think of subprograms as processes that (1) have been located outside the body of an architecture, and (2) operate only on their input and (in the case of procedures) their output parameters.

VHDL '93 Note:

In VHDL '93, functions can be declared as impure, *meaning that they can operate on and modify objects other than those passed as parameters. Impure functions can therefore have* side-effects, *and they can return different values for the same input parameters.*

Nesting of functions and procedures is allowed to any level of complexity, and recursion is also supported in the language. (Of course, if you expect to generate actual hardware from your VHDL descriptions using synthesis tools, then you will need to avoid writing recursive functions and procedures, as such descriptions are not synthesizable).

Declaring a Global Subprogram

Functions and procedures can be declared either globally, so they are usable throughout a design description, or they can be declared locally within the declarative region of an architecture, block, process, or even within another subprogram. If you are writing a subprogram that will be used throughout your design, you will write the subprogram declaration in an external package, as shown in the following example:

```
package my_package is
    function my_global_function(...)
        return bit;
end my_package;

package body my_package is
    function my_global_function(...)
        return bit is
    begin
        . . .
    end my_global_function;
end my_package;
. . .
use work.my_package.my_global_function;
entity my_design is
begin
    . . .
end my_design;
```

A global subprogram must be declared within a package.

In this example, the function **my_global_function()** has been declared within the package **my_package**. The actual body of the function—the sequence of statements that define its operation—is placed into a package body. (The reasons why a subprogram requires a package body in addition to a package are somewhat obscure, but they have to do with the fact that the statements in a subprogram must be executed when the design description is simulated, while other declarations appearing in a package can be completely resolved at the time the VHDL description is first analyzed by the VHDL compiler.) To use the global function in subsequent architectures (such as the architecture associated with entity **my_design** in this example), a **use** statement (and **library** statement, if the package has been compiled into a named library) must precede the declaration for that architecture or its parent entity.

Declaring a Local Subprogram

Another way of using subprograms is to declare them locally, such as within an architecture or block declaration. In the following example, **my_local_function()** has been declared entirely within the architecture **my_architecture**.

Local subprograms can be declared within architectures, processes, and blocks.

```
architecture my_architecture of my_design is
begin
   my_process: process(...)
      function my_local_function(...)
         return bit is
      begin
         . . .
      end my_local_function;
   begin
      . . .
   end process my_process;
end my_architecture;
```

This example demonstrates the concept of local scoping. We saw in Chapter 3, *Exploring Objects and Data Types*, that VHDL objects (such as signals, variables and constants) can be declared at many points in a design, and that the visibility, or scoping, of those objects depends on where they have been declared. Subprograms (functions and procedures) also have scoping. In this example, the function **my_local_function** can only be referenced within the architecture in which it has been declared and defined.

Consistent scoping of objects and subprograms is an important part of modular VHDL coding and of structured programming in general. If you will only be using an object or subprogram in one section of your overall design, then you should keep the declaration of that object or subprogram local to that section of the design. This will make it possible to re-

Functions

A function is a subprogram that accepts zero or more input arguments and returns a single output value. Because a function returns a value, it has a type associated with it. The following is an example of a function that accepts two integer arguments and returns the greater of the two as an integer value:

```
function maxval (arg1, arg2: integer) return integer is
    variable result: integer;
begin
    if arg1 > arg2 then
        result := arg1;
    else
        result := arg2;
    end if;
     return result;
end maxval;
```

A function returns a single value of a specified type.

The arguments to a function are all inputs to the function. They cannot be modified or otherwise assigned values within the function. By default, the arguments are of a *constant kind*. This means that the arguments are interpreted within the function as if they had been supplied as constants declared in the function itself. An alternative type of argument, indicated by the use of the **signal** keyword, allows the use of signal attributes (such as **'event**) within the function. The following function (which is provided in the IEEE 1164 standard library) demonstrates the use of a **signal** argument in a function:

```
function rising_edge (signal s: std_logic) return boolean is
begin
    return (s'event and (To_X01(s) = '1') and
        (To_X01(s'last_value) = '0'));
end;
```

In this example, the keyword **signal** is critical to the correct operation of the function. In the absence of the signal keyword, the **'event** attribute would not be preserved.

Functions are most commonly used in situations where you require a calculation or conversion based on the subprogram inputs. Examples of this include arithmetic or logic functions (such as the one just presented), type conversion functions, and value checks such as you might use when writing a test bench.

Because they return a value, functions must be used as part of a larger expression. The following VHDL code fragment demonstrates a type conversion function being used in an expression to convert an array data type to an integer:

```
signal Offset: integer range (0 to 1023);
signal BUS1: std_logic_vector(11 downto 0);

    . . .

Offset <= to_integer(BUS1) + 136;
```

Operators as Functions

Overloading makes it possible to write custom operations for differing data types.

One interesting feature of VHDL is its support for operator *overloading*. Operator overloading allows you to specify custom functions representing symbolic operations for your own data types. To define a new operation (or modify an existing one), you simply write a function and enclose its name (which can be a non-numeric name such as an operator symbol) in double-quote characters.

The following operator function is taken directly from the IEEE 1164 standard logic package, and demonstrates how operator overloading works:

```
function "and" (l : std_logic; r : std_logic ) return UX01 is
begin
    return(and_table(l, r));
end "and";
```

In this example, the function **and** is declared as a function returning the type **UX01** (a four-valued logic type used internally in the standard logic package). The function is identified during compilation by its name (**and**) and by the types and number of its arguments. For example, in the expression:

```
architecture simple of and_operation is
    signal Y, A, B: std_logic;
begin
    Y <= A and B;
end simple;
```

the **and** operation is actually a function defined using the previously listed statements. In fact, all of the standard operations that you use in VHDL (including such operators as **and**, **or**, **not**, **+**, **-**, *****, **&** and **<**) are actually functions declared in libraries such as **std** and **ieee**.

Note:
In source code listings presented in this book, we have used the typographic convention of listing all VHDL keywords in bold face. As you have just seen, however, many of the keywords that we list in bold face are actually functions defined in a standard library.

Subprogram Overloading

Functions and procedures are uniquely identified by the types of their arguments, as well as by their names.

Because a function or procedure is uniquely identified by its name in combination with its argument types, there can be more than one function or procedure defined with the same name, depending on the types of the operands required. This feature (called *subprogram overloading*) is important because the function required to perform a given operation on one type of data may be quite different than the function required for another type.

It is unlikely that you will need to use subprogram overloading in your own design efforts. Instead, you will use the standard data types provided for you in the language standards, and you will use the predefined operators for those data types exclusively. You might find it useful, however, to look over the operators defined in the standard libraries so you have a better idea of the capabilities of each standard data type provided.

Procedures

Procedures do not have return values...

Procedures differ from functions in that they do not have a return value, and their arguments may include both inputs and outputs to the subprogram. Because each argument to a procedure has a mode (**in**, **out**, or **inout**), they can be used very much like you would use an entity/architecture pair to help simplify and modularize a large and complex design description.

Procedures are used as independent statements, either within the concurrent area of an architecture or within the sequential statement area of a process or subprogram.

The following sample procedure defines the behavior of a clocked JK flip-flop with an asynchronous reset:

```
procedure jkff (signal Rst, Clk: in std_logic;
                signal J, K: in std_logic;
                signal Q,Qbar: inout std_logic) is
begin
  if Rst = '1' then
    Q <= '0';
  elsif Clk = '1' and Clk'event then
    if J = '1' and K = '1' then
      Q <= Qbar;
    elsif J = '1' and K = '0' then
      Q <= '1';
    elsif J = '0' and K = '1' then
      Q <= '0';
    end if;
  end if;
  Qbar <= not Q;
end jkff;
```

A procedure may include a **wait** statement, unless it has been called from within a process that has a sensitivity list.

Note that variables declared and used within a procedure are not preserved between different executions of the procedure. This is unlike a process, in which variables maintain their values between executions. Variables within a procedure therefore do not maintain their values over time, unless the procedure is suspended with a **wait** statement.

Parameter Types

Subprograms operate on values or objects that are passed in as parameters to the subprogram. Procedures differ from functions in that they can also pass information out on the parameter list. (The parameters of a procedure have directions, or *modes*.)

There are three classes of parameters available for subprograms....

There are three classes of parameters available for subprograms: **constant, variable** and **signal**. The default class, if no other class is specified, is **constant**. The parameters that are used within the function or procedure are called the *formal parameters*, while the parameters passed into the function or procedure are called the *actual parameters*.

The primary difference between **constant, variable** and **signal** parameters is the type of actual parameters that can be passed into the subprogram when it is called. If the formal parameter of a subprogram is of class **constant**, the actual parameter can be any expression that evaluates to a data type matching that of the formal parameter. For parameters of class **variable** or **signal**, the actual parameters must be **variable** or **signal** objects, respectively.

Parameters of subprograms transfer only the value of the actual parameters (those parameters specified when the subprogram is called) for the formal parameters (the parameters specified in the subprogram declaration). Attribute information is not passed directly into the subprogram. (The attributes that you will most often be concerned with, such as **'event**, will be available if you are using parameters of class **signal**.)

Mapping of Parameters

The examples presented in this chapter have used what is refered to as *positional association* to describe how actual parameters are paired with formal parameters of the subprogram.

Positional association is a quick and convenient way to describe the mapping of parameters, but it can be error-prone.

Subprogram parameters can be associated by position or by name.

For this reason, you might want to write your subprogram references using an alternate form of port map called *named association*. Named association guarantees that the correct parameters are connected, and it also gives you the ability to re-order the parameters as needed.

The following example shows how the same subprogram might be referenced using both positional and named association:

```
dff(Rst,Clk,Data,Result);
dff(Rst=>Rst,C=>Clk,D=>Data,a=>Result);
```

The special operator => indicates exactly which lower-level ports are to be connected to which higher-level signals.

Is There More?

Yes. VHDL includes other features to help modularize your designs and manage their complexity. We'll lead you through them in the next chapter as you begin...

8. Partitioning Your Design

VHDL provides many high-level features to help you manage a complex design description. In fact, design management is one of VHDL's key strengths when compared to alternative design entry languages and methods.

Design partitioning goes beyond simpler design modularity methods...

The modularity features (procedures and functions) that we have seen in previous chapters are one aspect of design management, allowing commonly-used declarations and sequential statements to be collected in one place. Design partitioning is another important aspect of design management. Design partitioning goes beyond simpler design modularity methods to provide comprehensive design management across multiple projects and allow alternative structural implementations to be tried out with minimal effort.

Design partitioning is particularly useful for those designs being developed in a team environment, as it promotes cooperative design efforts and well-defined system interfaces.

The design partitioning features of VHDL include:

- Blocks
- Packages

- Libraries
- Components
- Configurations

Blocks

Blocks allow the logical grouping of statements within an architecture.

Blocks are the simplest form of design partitioning. They provide an easy way to segment a large VHDL architecture into multiple self-contained parts. Blocks allow the logical grouping of statements within an architecture, and provide a place to declare locally-used signals, constants, and other objects as needed.

VHDL blocks are analogous to sheets in a multi-sheet schematic. They do not represent re-usable components (unless you re-use them by copying them with your text editor or by using configurations), but do enhance readability by allowing declarations of objects to be kept close to where those objects are actually used.

The general form of the block statement is shown below:

```
architecture my_arch of my_entity is
begin

    BLOCK1: block
       signal a,b: std_logic;
    begin
       -- some local statements here
    end block BLOCK1;

    BLOCK2: block
       signal a,b std_logic;
    begin
       -- some other local statements here
       -- Note that 'a' and 'b' are unique to this block!
    end block BLOCK2;

end my_arch;
```

This simple example includes two blocks, named **BLOCK1** and **BLOCK2**, that each include declarations for local signals. In the first block, **BLOCK1**, the signals **a** and **b** are declared prior to the **begin** statement of the block. These signals are therefore local to block **BLOCK1** and are not visible outside of it. The second block, **BLOCK2**, also has declarations for local signals named **a** and **b**, but these are not the same signals as those declared in block **BLOCK1**.

This concept of local declarations is important to understand and is probably familiar to you if you have used high-level programming languages. One of the most important techniques of structured programming (whether you are describing software or hardware) is to minimize the overall complexity of your design description by localizing the declarations as much as is practical. Keeping signals local will make the design description easier to read, allow it to be modified more easily in the future, and also enhance design re-use, since it will be easier to copy one portion of the design to another project or source file.

Nested Blocks

Blocks can be nested, as shown in the following example:

```
architecture my_arch of my_entity is
begin

    BLOCK1: block
        signal a,b: std_logic;
    begin

        BLOCK2: block
            signal c,d std_logic;
        begin
            -- This block is now local to block BLOCK1 and has
            -- access to 'a' and 'b'
        end block BLOCK2;

    end block BLOCK1;

end my_arch;
```

In this example, block **BLOCK2** has been placed within block **BLOCK1**. This means that all declarations made within **BLOCK1** (signals **a** and **b**, in this example) are visible both within block **BLOCK1** and block **BLOCK2**. The reverse is not true, however. The declarations for **c** and **d** within block **BLOCK2** are local only to **BLOCK2** and are not visible outside that block.

What happens when the same signals are declared in two blocks that are nested? Consider the following:

```
architecture my_arch of my_entity is
begin

    BLOCK1: block
        signal a,b: std_logic;
    begin

        BLOCK2: block
            signal a,b std_logic;
        begin
            -- This a and b overrides previous
        end block BLOCK2;

    end block BLOCK1;

end my_arch;
```

Using the same object names within nested blocks can be confusing.

In this example, the signals **a** and **b** are declared both in the outer block (**BLOCK1**) and in the inner block (**BLOCK2**). The result is that the signals **a** and **b** in the outer block are hidden (but not replaced or overwritten) by the declarations of **a** and **b** in the inner block.

Note:

If you need to access a signal that has been effectively hidden by a declaration of the same name, you can qualify the signal name with a block name prefix, as in **BLOCK1.a** *or* **BLOCK1.b**.

Guarded Blocks

Guarded blocks are used to enable and disable drivers using a guard expression.

Guarded blocks are special forms of block declarations that include an additional expression known as a *guard expression*. The guard expression enables or disables drivers within the block, allowing circuits such as latches and output enables to be easily described using a dataflow style of VHDL.

The following example shows how a guarded block can be used to described the operation of a latch:

```
use ieee.std_logic_1164.all;
entity latch is
    port( D, LE: in std_logic;
          Q, QBar: out std_logic);
end latch;

architecture mylatch of latch is
begin

    L1: block (LE = '1')
    begin
      Q <= guarded D after 5 ns;
      QBar <= guarded not(D) after 7 ns;
    end block L1;

end mylatch;
```

In this example, the guard expression **LE = '1'** applies to all signal assignments that include the **guarded** keyword. (Guard expressions are placed in parentheses after the **block** keyword.) The signal assignments for **Q** and **QBar** therefore depend on the value of **LE** being **'1'**. When **LE** is not **'1'**, the guarded signals hold their values.

Note:

Guarded blocks are not supported by all synthesis tools, so it is not recommended that you use them for designs intended for synthesis. Instead, you should use a process or subprogram to describe the behavior of registered or latched circuits. Chapter 6, Understanding Sequential Statements, discusses a number of ways of describing register and latch behavior.

Packages

Packages are often compiled into named libraries for easy re-use.

Packages are intended to hold commonly-used declarations such as constants, type declarations and global subprograms. Packages can be included within the same source file as other design units (such as entities and architectures) or may be placed in a separate source file and compiled into a named library. This latter method (described in a later section) is useful when you will be using the contents of a package throughout a large design or in multiple projects.

Packages may contain the following types of objects and declarations:

- Type and subtype declarations

- Constant declarations

- File and alias declarations

- Component declarations

- Attribute declarations

- Functions and procedures

- Shared variables (VHDL 1076-1993 only)

When items from the package are required in other design units, you must include a **use** statement to make the package and its contents visible for each design unit.

The following is an example of a package declaration and its corresponding **use** statements:

```
library ieee;
use ieee.std_logic_1164.all;
package my_types is
   subtype byte is std_logic(0 to 7);
   constant CLEAR: byte := (others=>'0');
end my_types;


use work.mytypes.all;
use ieee.std_logic_1164.all;
```

```
entity rotate is
    port(Clk, Rst, Load: in std_logic;
          Data: in byte;
          Q: out byte);

end rotate;

architecture rotate4 of rotate is
    signal Qreg: byte;
begin

    Qreg <= Data when (Load = '1') else
            Qreg(1 to byte'LENGTH-1) & Qreg(0);

    dff(Rst, Clk, Qreg, Q);

end rotate4;
```

In this example, the package **my_types** includes declarations for a subtype (**byte**) and constant (**CLEAR**) that will be used throughout the subsequent design description. The statement **use work.mytypes.all** specifies that all contents of the package **mytypes** should be loaded from the default library (**work**). (As we will see later in this chapter, the **work** library is a special library described in the VHDL specification as one that does not require a **library** statement and into which all design units are analyzed by default.) An alternative to using the **all** keyword in the **use** statement would be to specify precisely which items in the default library are to be made visible, as in **use work.mytypes.byte** and **use work.mytypes.CLEAR**.

How Are Packages Used?

When you create your own VHDL design descriptions, you can use packages in a number of ways. First, you can dramatically simplify your designs by placing commonly-used declarations (such as **byte** and **CLEAR** in the previous example) into packages that are used throughout your project. You will probably find that using libraries to collect such packages in one place will simplify the design even further and make it easier to share commonly-used declarations between different design descriptions.

Another way you can use packages is to reference pre-written packages that have been provided for you. One example of such a package is found in the IEEE 1164 standard. The IEEE 1164 standard provides a standard package named **std_logic_1164** that includes declarations for the types **std_logic**, **std_ulogic**, **std_logic_vector** and **std_ulogic_vector**, as well as many useful functions related to those data types.

Packages may also be provided to you by vendors of synthesis and simulation tools. Synthesis tools, for example, often include packages containing synthesizable type conversion functions, synthesizable procedures for flip-flops and latches, and other useful design elements.

Finally, there is a standard package that includes declarations for all the standard data types (**bit**, **bit_vector**, **integer** and so on). This standard package is defined by the IEEE 1076 standard and automatically made visible to all design units. (You do not have to specify a **use** clause for the standard package.)

Package Bodies

Package bodies are required for global subprograms and deferred constants.

Packages that include global subprograms (functions or procedures) or deferred constants (see Chapter 3, *Exploring Objects and Data Types*) must defer part of their declaration (the part that must be analyzed during simulation) to a separate design unit called a *package body*. Every package can have, at most, one corresponding package body. Package bodies are optional and are only required when a package includes subprograms or deferred constants.

The following example shows how a package body must be used when a subprogram (in this case, a procedure describing the behavior of a D flip-flop) is declared in a package:

package my_reg8 **is**

```
        subtype byte8 is std_logic_vector(0 to 7);
        constant CLEAR8: byte8 := (others=>'0');
        procedure dff8 (signal Rst, Clk: in std_logic;
                    signal D: in byte;
                    signal Q: out byte);
    end my_reg8;

    package body my_reg8 is
        procedure dff8 (signal Rst, Clk: in std_logic;
                    signal D: in byte8;
                    signal Q: out byte8) is
        begin
          if Rst = '1' then
             Q <= CLEAR8;
          elsif Clk = '1' and Clk'event then
             Q <= D;
          end if;
        end dff;
    end my_reg8;
```

In this example, the procedure **dff8** is declared initially in the package **my_reg8**. This first declaration is somewhat akin to a "function prototype" as used in the C or C++ languages, and it defines the interface to the procedure. The package body that corresponds to package **my_reg8** (and shares its name) contains the complete description of the procedure.

Design Libraries

The actual implementation of design libraries depends on the simulation or synthesis tool you are using.

A design library is defined in the VHDL 1076 standard as "an implementation-dependent storage facility for previously analyzed design units". This rather loose definition has resulted in many different implementations in synthesis and simulation tools. In general, however, you will find that design libraries are used to collect commonly-used design units (typically packages and package bodies) into uniquely-named areas that can be referenced from multiple source files in your design.

In a typical simulation environment, you will specify to the simulator the library into which you want each design unit compiled (or analyzed, to use the terminology of the VHDL standard). If you do not specify a library, the design units are compiled into a default library named **work**.

For simple design descriptions (such as those that are completely represented within a single source file), you will use the **work** library exclusively and will not have to put much thought into how libraries are implemented in the set of tools you are using. When you use the **work** library exclusively, all you need to do is specify a **use** statement such as:

*You must write a **use** statement to make the contents of a package visible.*

use work.my_package.**all**;

prior to each entity declaration in your design for each package that you have declared in your source file. (You do not have to place **use** statements prior to an architecture declaration if the corresponding entity declaration is preceded by a **use** statement.)

If, however, you choose to use named libraries in your designs (and you are encouraged to do so, as it can dramatically improve your design productivity), then you should follow a few simple rules to avoid compatibility problems when moving between different simulation and synthesis environments. First, you should not use the **work** library to contain packages that are shared between design units located in different source files. Although some simulation environments allow previously-compiled contents of the **work** library to be accessed at any time (such as during the separate compilation of a source file), this is not actually defined by the VHDL standard and may not work in other simulation and synthesis environments.

Some synthesis and simulation tools actually define the **work** library to be only those design units that are included in the source file currently being compiled. This is a simple rule of usage and is the recommended use of **work**.

To keep your use of libraries as simple as possible, it is recommended that you make consistent use of VHDL source file names and corresponding library file names, and avoid the use of **work** for all but the simplest packages. The example of Figure 8-1 demonstrates how the use of libraries can simplify a design description:

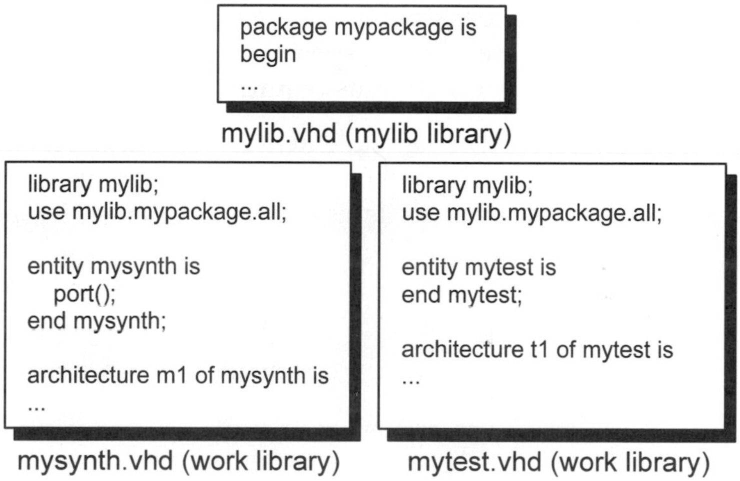

```
package mypackage is
begin
...
```

mylib.vhd (mylib library)

Figure 8-1: You can simplify your design and ensure portability by always compiling shared packages into named libraries, rather than into the **work** library.

```
library mylib;
use mylib.mypackage.all;

entity mysynth is
    port();
end mysynth;

architecture m1 of mysynth is
...
```

mysynth.vhd (work library)

```
library mylib;
use mylib.mypackage.all;

entity mytest is
end mytest;

architecture t1 of mytest is
...
```

mytest.vhd (work library)

In this example, the source file **mylib.vhd** includes a package named **mypackage**. The **mylib.vhd** source file will be compiled into a library named **mylib**, which is then referenced within both the synthesizable module represented by source file **mysynth.vhd** and in the test bench represented by source file **mytest.vhd**. Note that both **mysynth.vhd** and **mytest.vhd** are compiled into the default library (**work**), but there is no explicit reference (**library** statement) needed between them. The result of structuring the design in this way is that **work** is never referenced in a **library** or **use** statement, and the design is guaranteed to be compatible with any IEEE standard 1076-compliant VHDL simulation or synthesis environment, regardless of how libraries have been implemented by the tool vendor.

Package Visibility

*Packages are not visible unless the appropriate **library** and **use** statements have been included.*

The **library** statement described in the previous section is used to load a library so that its contents are available when compiling a source file. However, the **library** statement does not actually make the contents of the specified library (the packages or other design units found in the library) *visible* to design units in the current source file. Visibility is created when you specify one or more **use** statements prior to the design units requiring access to items in the library.

The **use** statement is quite flexible. You can specify exactly which items within a package are to be made visible, specify that all items in a package are to be made visible, or specify that all items in all packages for a specific library are to be made visible. The following examples demonstrate some of the possible uses of **use** statements:

use mylib.my_package.**all**;   -- All items in my_package are visible

use mylib.my_package.dff;   -- Just using the dff procedure

use mylib.**all**;   -- Make everything in the library visible

In general, you will find that it is most convenient to place a **library** statement (one for each external library being used) at the beginning of your source file, and place **use** statements just prior to those design units requiring visibility of items in the library. To prevent compatibility problems as described above, you should avoid using **work** for shared packages or other design units that cross source file boundaries.

For clarity, it is recommended that you specify both the library and package name in your **use** statements, even if you are using all items in the library.

Components

As we saw in Chapter 2, *A First Look At VHDL*, components are used to connect multiple VHDL design units (entity/architecture pairs) together to form a larger, hierarchical design. Using hierarchy can dramatically simplify your design description and can make it much easier to re-use portions of the design in other projects. Components are also useful when you want to make use of third-party design units, such as simulation models for standard parts, or synthesizable core models obtained from a company specializing in such models.

The following example describes, once again, the relationship between the three design units in our earlier shift and compare example:

Figure 8-2: The ***shiftcomp*** *design description includes references to two lower-level components,* ***shift*** *and* ***compare***.

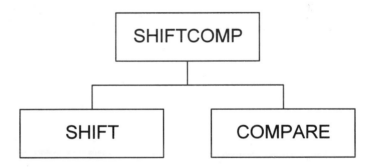

```
architecture structure of shiftcomp is

    component compare
        port(A, B: in bit_vector(0 to 7); EQ: out bit);
    end component;

    component shift
        port(Clk, Rst, Load: in bit;
            Data: in bit_vector(0 to 7);
            Q: out bit_vector(0 to 7));
    end component;

    signal Q: bit_vector(0 to 7);

begin
```

```
COMP1: compare port map (Q, Test, Limit);
SHIFT1: shift port map (Clk, Rst, Load, Init, Q);

end structure;
```

In this example, the two lower-level components (**shift** and **compare**) were *instantiated* in the higher-level module (**shiftcomp**) to form a hierarchy of design units. Each *component instantiation* is represented by a component name that is unique within the architecture or block.

Component instantiations are concurrent statements and therefore have no order-dependency. A design unit (such as this one) that includes only component instantiation statements can be thought of as a *netlist*, such as might be written (or generated) to represent the connections on a schematic.

Mapping of Ports

The previous example of component instantiation used *positional association* to describe how signals at the higher level (in this case **shiftcomp**) are to be matched with (i.e., connected to) ports of the entities in the lower-level modules (**shift** and **compare**).

Positional association is a quick and convenient way to describe the mapping of signals to ports in a component instantiation, but it can be error-prone. Consider, for example, what would have happened if the component instantiation for the **shift** module had been written as follows:

Positional association can make connection errors difficult to detect.

```
SHIFT1: shift port map (Rst, Clk, Load, Init, Q);
```

Because the **Rst** and **Clk** signals are of the same type (**std_logic**), the simulator or synthesis tool would accept this port mapping without complaint, and it would connect the reset signal to the clock and connect the clock to the reset. The circuit would not operate as expected, and the problem might be difficult to debug.

For this reason, we generally recommend that you write component instantiations using an alternate form of port map called *named association*. Named association guarantees that the correct signals and ports are connected through the hierarchy, and it also gives you the ability to re-order the ports as needed.

The following example shows how the same component (a NAND gate) might be instanced using both positional and named association:

```
U1: nand2 port map (a, b, y);                  -- Positional association
U2: nand2 port map (a=>in1,b=>in2,y=>out1);    -- Named association
```

The special operator => indicates exactly which lower-level ports (**a**, **b** and **y**, in this case) are to be connected to which higher-level signals (**in1**, **in2** and **out1**).

Named association also makes it possible to leave one or more lower-level ports unconnected using the keyword **open**, as shown below:

Named association can be more clear and less error-prone.

```
U2: count8 port map (C => Clk1, Rst => Clr, L => Load, D => Data,
                     Q => , Cin => open);
```

Note:
You might also consider placing each named association on a separate line. This simplifies debugging because the debugger will identify the exact line where an association error occurred.

Generics

Generics make it possible to pass instance-specific information to a component.

It is possible to pass instance-specific information other than actual port connections to an entity using a feature called generics. Generics are very useful for making design units more general-purpose or for annotating information (such as timing specifications) to an entity at the time the design is analyzed for simulation or synthesis.

The following example shows how generics can be used to create a parameterizable model of a D-type flip-flop:

```
library ieee;
use ieee.std_logic_1164.all;

entity dffr is
   generic (wid: positive);
   port (Rst,Clk: in std_logic;
        signal D: in std_logic_vector(wid-1 downto 0);
        signal Q: out std_logic_vector(wid-1 downto 0));
end dffr;

architecture behavior of dffr is
begin
   process(Rst,Clk)
      variable Qreg: std_logic_vector(wid-1 downto 0);
   begin
      if Rst = '1' then
         Qreg := (others => '0');
      elsif Clk = '1' and Clk'event then
         for i in Qreg'range loop
            Qreg(i) := D(i);
         end loop;
      end if;
      Q <= Qreg;
   end process;
end behavior;
```

Generics can be used to parameterize an entity, making it more general-purpose.

In this example, the **dffr** entity has a generic list in addition to a port list. This generic list contains one entry, a positive integer, that corresponds to the width of the **D** input and **Q** output. The architecture declaration uses a **for** loop in conjunction with the generic (**wid**) to describe the operation of the D-type flip-flops.

When instantiated in a higher-level design unit, a generic map must be provided in addition to the port map, as shown below:

```
architecture sample of reg is

   component dffr
      generic (wid: positive);
      port (Rst,Clk: in std_logic;
           signal D: in std_logic_vector(wid-1 downto 0);
           signal Q: out std_logic_vector(wid-1 downto 0));
   end component;
```

```
    constant WID8: positive := 8;
    constant WID16: positive := 16;
    constant WID32: positive := 32;
    signal D8,Q8: std_logic_vector(7 downto 0);
    signal D16,Q16: std_logic_vector(15 downto 0);
    signal D32,Q32: std_logic_vector(31 downto 0);

begin

    FF8:  dffr generic map(WID8)  port map(Rst,Clk,D8,Q8);
    FF16: dffr generic map(WID16) port map(Rst,Clk,D16,Q16);
    FF32: dffr generic map(WID32) port map(Rst,Clk,D32,Q32);

end sample;
```

The example shows how three instances of the **dffr** design unit can be created using different values for the generic.

Configurations

Configurations allow large, complex design descriptions to be better managed.

Configurations are features of VHDL that allow large, complex design descriptions to be managed during simulation. (Configurations are not generally supported in synthesis.) One example of how you might use configurations is to construct two versions of a system-level design, one of which makes use of high-level behavioral descriptions of the system components, while a second version substitutes in a post-synthesis timing model of one or more components.

A *configuration declaration* is a primary design unit that defines the binding of some or all of the component instances in your design description to corresponding lower-level entities and architectures. The configuration declaration can form a simple parts list for your design, or it can be written to contain detailed information about how each component is "wired into" the rest of the design (through specific port mappings) and the values for generics being passed into each entity.

If you think of the configuration declaration as a parts list for your design, you can perhaps visualize it better as follows: consider a design description in which you have described an entity named **Board** with an architecture named **structure**. In the architecture structure you have described one instance (**U1**) of a component called **Chip**. Moving down in the hierarchy of your design, let's suppose that the entity **Chip** has been written with four alternative architectures named **A1, A2, A3** and **A4**. (There are many reasons why you might have done this. For example, the default architecture might be the final synthesizable version of the chip, while the remaining three are versions intended strictly for high-level simulation.)

One View of Configurations

Figure 8-3: Consider an entity with multiple alternative architectures.

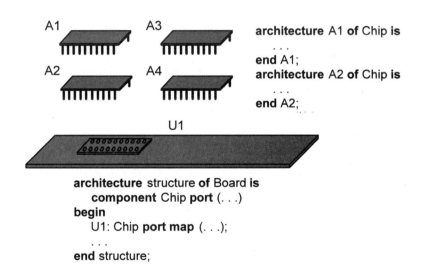

```
architecture A1 of Chip is
    . . .
end A1;
architecture A2 of Chip is
    . . .
end A2;
```

```
architecture structure of Board is
    component Chip port (. . .)
begin
    U1: Chip port map (. . .);
    . . .
end structure;
```

The purpose of a configuration declaration in this case might be to select which of the four alternate architectures is to be used in the current simulation, as follows:

Figure 8-4: A configuration declaration might be used to select a specific architecture for simulation.

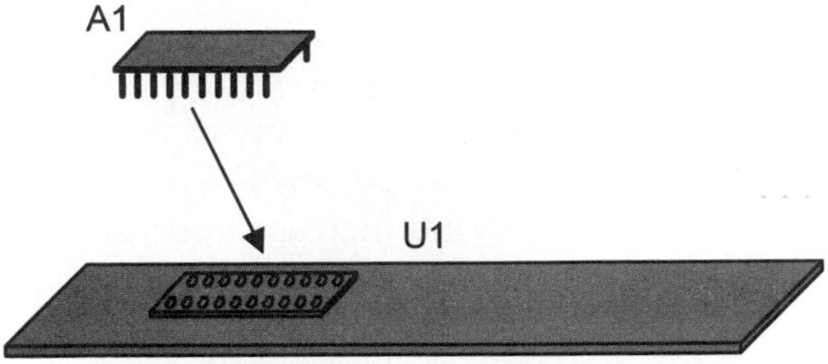

```
configuration this_build of Board is
    for structure
        for U1: Chip
            use entity work.Chip(A1);
            port map(reset => gnd, cin => 1 . . .);
        end for;
    end for;
end this_build;
```

There are many applications of configurations in simulation. For large projects involving many engineers and many design revisions, configurations can be used to manage versions and specify how a design is to be configured for system simulation, detailed timing simulation, and synthesis. Because simulation tools allow configurations to be modified and recompiled without the need to recompile other design units, it is easy to construct alternate configurations of a design very quickly without having to recompile the entire design.

Configurations are not generally supported in synthesis tools.

Because configurations are not generally supported in synthesis tools, we will not describe the many advanced uses of configurations for simulation and design management. Instead, you are encouraged to read about them in the VHDL Language Reference Manual (IEEE Standard 1076) or in a book more oriented toward simulation modeling.

So What's Left?

We've explored many aspects of VHDL. We've seen how you can use this powerful language to streamline your circuit designs and increase your productivity. Yet VHDL's usefulness goes well beyond just designing circuits themselves. There's much more; let's see how we can use VHDL for...

9. Writing Test Benches

As you become proficient with simulation, you will find that your VHDL simulator becomes your primary design tool.

While much of this book has focused on the uses of VHDL for synthesis, one of the primary reasons to use VHDL is its power as a test stimulus language. As logic designs become more complex, comprehensive, up-front verification becomes critical to the success of a design project. In fact, as you become proficient with simulation, you will quickly find that your VHDL simulator becomes your primary design development tool. When simulation is used right at the start of the project, you will have a much easier time with synthesis, and you will spend far less time re-running time-intensive processes, such as FPGA place-and-route tools and other synthesis-related software.

To simulate your project, you will need to develop an additional VHDL program called a test bench. (Some VHDL simulators include a command line stimulus language, but these features are no replacement for a true test bench.) Test benches emulate a hardware breadboard into which you will "install" your synthesizable design description for the purpose of verification. Test benches can be quite simple, applying a sequence of inputs to the circuit over time. They can also be quite complex, perhaps even reading test data from a disk file and writing test results to the screen and to a report file. A comprehensive test bench can, in fact, be more complex and lengthy (and take longer to develop) than the synthesizable circuit being tested. As you will begin to appreciate while reading this chapter, test bench development will be where you make use of the full power of VHDL and your own skills as a VHDL "coder".

Depending on your needs (and whether timing information related to your target device technology is available), you may develop one or more test benches to verify the design functionally (with no delays), to check your assumptions about timing relationships (using estimates or unit delays), or to simulate with annotated post-route timing information so you can verify that your circuit will operate in-system at speed.

During simulation, the test bench will be the top level of a design hierarchy. To the simulator, there is no distinction between those parts of the design that are being tested and the test bench itself. In your own mind, however, you can think of the test bench as a separate circuit, analogous to a large automated tester (Figure 9-1).

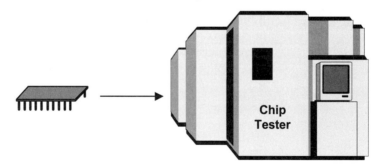

Figure 9-1: *A test bench can be thought of as a "virtual tester" into which you plug your design for verification.*

In most of this book, we have been emphasizing those aspects of the VHDL language that are synthesizable. In doing so, we have actually seen only a subset of the VHDL language in the examples presented. When writing test benches, you will most likely use a broader range of language features. You may use records and multi-dimensional arrays to describe test stimuli, write loops, create subprograms to simplify repetitive actions, and/or use VHDL's text I/O features to read and write files of data.

A Simple Test Bench

The simplest test benches apply some sequence of inputs to the circuit...

The simplest test benches are those that apply some sequence of inputs to the circuit being tested (the *Unit Under Test*, or *UUT*) so that its operation can be observed in simulation. Waveforms are typically used to represent the values of signals in the design at various points in time. Such a test bench must consist of a component declaration corresponding to the unit under test, and a description of the input stimulus being applied to the UUT.

The following example demonstrates the simplest form of a test bench, and tests the operation of a NAND gate:

```
library ieee;              -- Load the ieee 1164 library
use ieee.std_logic_1164.all;  -- Make the package 'visible'

use work.nandgate;         -- We'll use the NAND gate model from 'work'

-- The top level entity of the test bench has no ports...
--
entity testnand is
end testnand;

architecture stimulus of testnand is
    -- First, declare the lower-level entity...
    component nand
      port  (A,B: in std_logic;
             Y: out std_logic);
    end component;

    -- Next, declare some local signals to assign values to and observe...
    signal A,B: std_logic;
    signal Y: std_logic;

begin
    -- Create an instance of the comparator circuit...
    NAND1: nandgate port map(A => A,B => B,Y => Y);
```

```
-- Now define a process to apply some stimulus over time...
process
    constant PERIOD: time := 40 ns;
begin
    A <= '1';
    B <= '1';
    wait for PERIOD;
    assert (Y = '0')
        report "Test failed!" severity ERROR;
    A <= '1';
    B <= '0';
    wait for PERIOD;
        assert (Y = '1')
        report "Test failed!" severity ERROR;
    A <= '0';
    B <= '1';
    wait for PERIOD;
        assert (Y = '1')
        report "Test failed!" severity ERROR;
    A <= '0';
    B <= '0';
    wait for PERIOD;
        assert (Y = '1')
        report "Test failed!" severity ERROR;
    wait;
  end process;
end stimulus;
```

Reading from the top of this test bench, we see:

- **Library** and **use** statements making the standard logic package available for use (our lower-level NAND gate model has been described using standard logic).

- An optional **use** statement referencing the lower-level design unit **nand** from the **work** library.

- An entity declaration for the test bench. Note that test benches do not generally include an interface (port) list, as they are the highest-level design unit when simulated.

- An architecture declaration, containing:

- A component declaration corresponding to the unit under test.

- Signal declarations for **A**, **B**, and **Y**. These local signals will be used to (1) apply inputs to the unit under test, and (2) observe the behavior or the output during simulation.

- A component instantiation statement and corresponding port map statement that associates the top-level signals **A**, **B** and **Y** with their equivalent ports in the lower-level entity. Note that the component name used (**UUT**) is not significant; any valid component name could have been chosen.

- A **process** statement describing the inputs to the circuit over time. This process has been written without the use of a sensitivity list. It uses **wait** statements to provide a specific amount of delay (defined using constant **PERIOD**) between each new combination of inputs. **Assert** statements are used to verify that the circuit is operating correctly for each combination of inputs. Finally, a **wait** statement without any condition expression is used to suspend simulation indefinitely after the desired inputs have been applied. (In the absence of the final **wait** statement, the process would repeat forever, or for as long as the simulator had been instructed to run.)

Using Assert Statements

Assert statements provide a quick and easy way to check expected values...

VHDL's **assert** statement provides a quick and easy way to check expected values and display messages from your test bench. An **assert** statement has the following general format:

assert *condition_expression*
    report *text_string*
    severity *severity_level* ;

233

When analyzed (either during execution as a sequential statement, or during simulator initialization in the case of a concurrent **assert** statement), the condition expression is evaluated. As in an **if** statement, the condition expression of an **assert** statement must evaluate to a boolean (**true** or **false**) value. If the condition expression is **false** (indicating the assertion *failed*), the text that you have specified in the optional **report** statement clause is displayed in your simulator's transcript (or other) window. The **severity** statement clause then indicates to the simulator what action (if any) should be taken in response to the assertion failure (or *assertion violation*, to use the language of the VHDL specification).

The severity level can be specified using one of the following predefined severity levels: **NOTE, WARNING, ERROR,** or **FAILURE**. The actions that result from the use of these severity levels will depend on the simulator you are using, but you can generally expect the simulator to display a file name and line number associated with the **assert** statement, keep track of the number of assertion failures, and print a summary at the end of the simulation run. **Assert** statements that specify **FAILURE** in their severity statement clauses will normally result in the simulator halting.

Displaying Complex Strings in Assert Statements

VHDL's built-in support for formatted strings is somewhat limited...

A common use of **assert** and **report** statements is to display information about signals or variables dynamically during a simulation run. Unfortunately, VHDL's built-in support for this is somewhat limited. The problem is twofold: first, the **report** clause only accepts a single string as its argument, so it is necessary to either write multiple **assert** statements to output multiple lines of information (as when formatting and displaying a table), or you must make use of the string concatenation operator **&** and the special character constant **CR** (carriage return) and/or **LF** (line feed) to describe a single, multi-line string as shown below:

```
assert false
    report "This is the first line of the message." & CR & LF &
        "This is the second line of the message.";
```

The second, more serious limitation of the **report** statement clause is that it only accepts a string, and there is no built-in provision for formatting various types of data (such as arrays, integers and the like) for display. This means that to display such data in an **assert** statement, you must provide type conversion functions that will convert from the data types you are using to a formatted string. The following example (which is described in more detail later in this chapter) demonstrates how you might write a conversion function to display a **std_logic_vector** array value as a string of characters:

A conversion function can be used to display an array as a formatted string.

```
architecture stimulus of testfib is
    . . .
    function vec2str(vec: std_logic_vector) return string is
    variable stmp: string(vec'left+1 downto 1);
    begin
        for i in vec'reverse_range loop
            if (vec(i) = 'U') then
                stmp(i+1) := 'U';
            elsif (vec(i) = 'X') then
                stmp(i+1) := 'X';
            elsif (vec(i) = '0') then
                stmp(i+1) := '0';
            elsif (vec(i) = '1') then
                stmp(i+1) := '1';
            elsif (vec(i) = 'Z') then
                stmp(i+1) := 'Z';
            elsif (vec(i) = 'W') then
                stmp(i+1) := 'W';
            elsif (vec(i) = 'L') then
                stmp(i+1) := 'L';
            elsif (vec(i) = 'H') then
                stmp(i+1) := 'H';
            else
                stmp(i+1) := '-';
            end if;
        end loop;
    return stmp;
    end;
    . . .
```

```
    signal S: std_logic_vector(15 downto 0);
    signal S_expected: std_logic_vector(15 downto 0);
begin
    . . .
    process
    begin
      . . .

    assert (S /= S_expected)   -- report an error if different
        report "Vector failure!" & CR & LF &
        "Expected S to be  " & vec2str(S_expected) & CR & LF &
        "but its value was " & vec2str(S)
        severity ERROR;
```

In this example, a type conversion function has been written (**vec2str**) that converts an object of type **std_logic_vector** to a string of the appropriate format and size for display. As you develop more advanced test benches, you will probably find it useful to collect such type conversion functions into a library for use in future test benches.

As we will see later in this chapter, there are other, more powerful ways to display formatted output, using the built-in text I/O features of the language.

Using Loops and Multiple Processes

Loops can dramatically simplify a test bench.

Test benches can be dramatically simplified through the use of loops, constants and other more advanced features of VHDL. Using multiple concurrent processes in combination with loops can result in very concise descriptions of complex input and expected output conditions.

The following example demonstrates how a loop (in this case a **while** loop) might be used to create a background clock in one process, while other loops (in this case **for** loops) are used to apply inputs and monitor outputs over potentially long periods of time:

```
Clock1: process
    variable clktmp: std_logic := '1';
begin
```

```vhdl
    while done /= true loop
        wait for PERIOD/2;
        clktmp := not clktmp;
        Clk <= clktmp;
    end loop;
    wait;
end process;

Stimulus1: Process
Begin
    Reset <= '1';
    wait for PERIOD;
    Reset <= '0';
    Mode <= '0';
    wait for PERIOD;
    Data <= (others => '1');
    wait for PERIOD;
    Mode <= '1';

    -- Check to make sure we detect the vertical sync...
    Data <= (others => '0');
    for i in 0 to 127 loop
        wait for PERIOD;
        assert (VS = '1')
            report "VS went high at the wrong place!" severity ERROR;
    end loop;
    assert (VS = '1')
        report "VS was not detected!" severity ERROR;

    -- Load in the test counter value to check the end of frame detection...
    TestLoad <= '1';
    wait for PERIOD;
    TestLoad <= '0';
    for i in 0 to 300 loop
        Data <= RandomData();
        wait for PERIOD;
    end loop;
    assert (EOF = '1')
        report "EOF was not detected!" severity ERROR;

    done <= true;
    wait;

End Process;

End stimulus;
```

In this example, the process labeled **Clock1** uses a local variable (**clktmp**) to describe a repeating clock with a period defined by the constant **PERIOD**. This clock is described with a **while** loop statement, and it runs independent of all other processes in the test bench until the **done** signal is asserted **true**. The second process, **Stimulus1**, describes a sequence of inputs to be applied to the unit under test. It also makes use of loops—in this case **for** loops—to describe lengthy repeating stimuli and expected value checks.

Writing Test Vectors

Another approach to creating test stimuli is to describe the test bench in terms of a sequence of fixed input and expected output values. This sequence of values (sometimes called *test vectors*) could be described using multi-dimensional arrays or using arrays of records. The following example makes use of a record data type, **test_record**, which consists of the record elements **CE**, **Set**, **Din** and **CRC_Sum**. An array type (**test_array**) is then declared, representing an unconstrained array of **test_record** type objects. The constant **test_vectors**, of type **test_array**, is declared and assigned values corresponding to the inputs and expected output for each desired test vector.

Test vectors are sequences of inputs and corresponding outputs expressed in tabular form.

The test bench operation is described using a **for** loop within a process. This **for** loop applies the input values **Set** and **Din** (from the test record corresponding to the current iteration of the loop) to the unit under test. (The **CE** input is used within the test bench to enable or disable the clock, and is not passed into the unit under test.) After a certain amount of time has elapsed (as indicated by a **wait** statement), the **CRC_Sum** record element is compared against the corresponding output of the unit under test, using an **assert** statement.

```
library ieee;
use ieee.std_logic_1164.all;

use work.crc8s;    -- Get the design out of library 'work'
```

```
entity testcrc is
end testcrc;

architecture stimulus of testcrc is
    component crc8s
        port (Clk,Set,Din: in std_logic;
            CRC_Sum: out std_logic_vector(15 downto 0));
    end component;

    signal CE: std_logic;
    signal Clk,Set: std_logic;
    signal Din: std_logic;
    signal CRC_Sum: std_logic_vector(15 downto 0);
    signal vector_cnt: integer := 1;
    signal error_flag: std_logic := '0';
```

Record data types can be useful for representing test vector data.

```
    type test_record is record           -- Declare a record type
        CE: std_logic;                   -- Clock enable
        Set: std_logic;                  -- Register preset signal
        Din: std_logic;                  -- Serial Data input
        CRC_Sum: std_logic_vector (15 downto 0);   -- Expected result
    end record;

    type test_array is array(positive range <>) of test_record;  -- Collect them
                                                                 -- in an array

    -- The following constant declaration describes the test vectors to be
    -- applied to the design during simulation, and the expected result after a
    -- rising clock edge.
    constant test_vectors : test_array := (
        -- CE, Set, Din, CRC_Sum
        ('0', '1', '0', "----------------"),  -- Reset

        ('1', '0', '0', "----------------"), -- 'H'
        ('1', '0', '1', "----------------"),
        ('1', '0', '0', "----------------"),
        ('1', '0', '0', "----------------"),
        ('1', '0', '1', "----------------"),
        ('1', '0', '0', "----------------"),
        ('1', '0', '0', "----------------"),
        ('1', '0', '0', "0010100000111100"), -- x283C

        ('1', '0', '0', "----------------"), -- 'e'
        ('1', '0', '1', "----------------"),
        ('1', '0', '1', "----------------"),
        ('1', '0', '0', "----------------"),
```

```
               ('1', '0', '0', "----------------"),
               ('1', '0', '1', "----------------"),
               ('1', '0', '0', "----------------"),
               ('1', '0', '1', "1010010101101001"), -- xA569

               ('1', '0', '0', "----------------"), -- 'l'
               ('1', '0', '1', "----------------"),
               ('1', '0', '1', "----------------"),
               ('1', '0', '0', "----------------"),
               ('1', '0', '1', "----------------"),
               ('1', '0', '1', "----------------"),
               ('1', '0', '0', "----------------"),
               ('1', '0', '0', "0010000101100101"), -- x2165

               ('1', '0', '0', "----------------"), -- 'l'
               ('1', '0', '1', "----------------"),
               ('1', '0', '1', "----------------"),
               ('1', '0', '0', "----------------"),
               ('1', '0', '1', "----------------"),
               ('1', '0', '1', "----------------"),
               ('1', '0', '0', "----------------"),
               ('1', '0', '0', "1111110001101001"), -- xFC69

               ('1', '0', '0', "----------------"), -- 'o'
               ('1', '0', '1', "----------------"),
               ('1', '0', '1', "----------------"),
               ('1', '0', '0', "----------------"),
               ('1', '0', '1', "----------------"),
               ('1', '0', '1', "----------------"),
               ('1', '0', '1', "----------------"),
               ('1', '0', '1', "1101101011011010")  -- xDADA
          );

begin
    -- instantiate the component
    UUT: crc8s port map(Clk,Set,Din,CRC_Sum);

    -- provide stimulus and check the result

    testrun: process
        variable vector : test_record;
    begin
        for index in test_vectors'range loop
            vector_cnt <= index;
            vector := test_vectors(index);    -- Get the current test vector
                        -- Apply the input stimulus...
```

```
        CE <= vector.CE;
        Set <= vector.Set;
        Din <= vector.Din;

               -- Clock (low-high-low) with a 100 ns cycle...
        Clk <= '0';
        wait for 25 ns;
        if CE = '1' then
           Clk <= '1';
        end if;
        wait for 50 ns;
        Clk <= '0';
        wait for 25 ns;

               -- Check the results...
        if (vector.CRC_Sum /= "----------------"
              and CRC_Sum  /= vector.CRC_Sum) then
           error_flag <= '1';
           assert false
              report "Output did not match!"
              severity WARNING;
        else
           error_flag <= '0';
        end if;
      end loop;
     wait;
    end process;
  end stimulus;
```

Note:

*VHDL 1076-1993 broadens the scope of bit string literals somewhat, making it possible to enter **std_logic_vector** data in non-binary forms, as in the constant hexadecimal value **x"283C"**.*

Reading and Writing Files with Text I/O

Text I/O features allow you to read and write files in your test bench.

The text I/O features of VHDL make it possible to open one or more data files, read lines from those files, and parse the lines to form individual data elements, such as elements in an array or record. To support the use of files, VHDL has the concept of a **file** data type, and includes standard, built-in functions for opening, reading from, and writing to file data types. (These data types and functions were described in Chapter 3, *Exploring Objects and Data Types*.) The **textio** package, which is included in the standard library, expands on the built-in file type features by adding text parsing and formatting functions, functions and special file types for use with interactive ("std_input" and "std_output") I/O operations, and other extensions.

Text I/O is one area in which the 1076-1987 and 1076-1993 language specifications differ. The file and text I/O features of VHDL changed in the 1076-1993 standard, making it necessary to explicitly open a file before reading data from it. We will show how file I/O can be described using both the 1987 and 1993 language standards.

VHDL 1076-1987 File I/O

The following example demonstrates how you can use the text I/O features of VHDL to read test data from an ASCII file, using the VHDL 1076-1987 standard text I/O features.

This test bench reads lines from an ASCII file and applies the data contained in each line as a test vector to stimulate and test a simple Fibonacci sequence generator circuit. It begins with our by-now-familiar entity-architecture pair:

*To use the text I/O features of the standard library, you must include a **use** reference to package **std.textio**.*

```
--------------------------------------------------------------
-- Test bench for Fibonacci sequence generator.
--
library ieee;
use ieee.std_logic_1164.all;
use std.textio.all;          -- Use the text I/O features of the standard library
use work.fib;                -- Get the design out of library 'work'
entity testfib is            -- Entity; once again we have no ports
end testfib;

architecture stimulus of testfib is
  component fib              -- Create one instance of the fib design unit
    port (Clk,Clr: in std_logic;
        Load: in std_logic;
        Data_in: in std_logic_vector(15 downto 0);
        S: out std_logic_vector(15 downto 0));
  end component;
```

```vhdl
-- Define some local conversion functions. These will be used to:
--    (a) convert strings of characters read from the file into arrays
--        of values
--    (b) convert arrays of std_logic to strings for display purposes
--
function str2vec(str: string) return std_logic_vector is
variable vtmp: std_logic_vector(str'range);
begin
   for i in str'range loop
      if (str(i) = '1') then
         vtmp(i) := '1';
      elsif (str(i) = '0') then
         vtmp(i) := '0';
      else
         vtmp(i) := 'X';
      end if;
   end loop;
   return vtmp;
end;

function vec2str(vec: std_logic_vector) return string is
variable stmp: string(vec'left+1 downto 1);
begin
   for i in vec'reverse_range loop
      if (vec(i) = '1') then
         stmp(i+1) := '1';
      elsif (vec(i) = '0') then
         stmp(i+1) := '0';
      else
         stmp(i+1) := 'X';
      end if;
   end loop;
return stmp;
end;

signal Clk,Clr: std_logic;              -- Declares local signals
signal Load: std_logic;
signal Data_in: std_logic_vector(15 downto 0);
signal S: std_logic_vector(15 downto 0);
signal done: std_logic := '0';

constant PERIOD: time := 50 ns;

for UUT: fib use entity work.fib(behavior);          -- Configuration
                                                     -- specification
```

```
      begin
         UUT: fib port map(Clk=>Clk,Clr=>Clr,Load=>Load,     -- Creates one
                       Data_in=>Data_in,S=>S);               -- instance

         Clock: process
            variable c: std_logic := '0';        -- Clock process, just like the last one
         begin
            while (done = '0') loop              -- The done flag indicates that we
               wait for PERIOD/2;                -- are finished and can stop the clock.
               c := not c;
               Clk <= c;
            end loop;
         end process;

         -- Process 'read_input' loops through all the lines in the file and
         -- extracts the test vector data. This process makes use of functions
         -- in the text I/O library (provided with IEEE Standard 1076) as well as
         -- the two conversion functions declared earlier in this architecture.
         -- Each line of the test vector file contains two fields: the input vector
         -- and the expected output values.
         Read_input: process
            file vector_file: text is in "testfib.vec";    -- Declare and open the
                                                           -- file (1076-1987 style).
            variable stimulus_in: std_logic_vector(33 downto 0);  -- Inputs
            variable S_expected: std_logic_vector(15 downto 0);  -- Outputs
            variable str_stimulus_in: string(34 downto 1);    -- The vector string
            variable err_cnt: integer := 0;                    -- Error counter
            variable file_line: line;                   -- Text line buffer; 'line' is a
                                                        -- standard type (textio library).

         begin
            wait until rising_edge(Clk);         -- Synchronize with first clock
            while not endfile(vector_file) loop      -- Loop through lines in the file

               readline (vector_file,file_line);     -- Read one complete line
                                                     -- into file_line.
               read (file_line,str_stimulus_in) ;    -- Extract the first field from
                                                     -- file_line.
               stimulus_in := str2vec(str_stimulus_in);   -- Convert the input
                                                          -- string to a vector

               wait for 1 ns;                        -- Delay for a nanosecond

               Clr <= stimulus_in(33);               -- Get each input's
               Load <= stimulus_in(32);              -- value from the test
               Data_in <= stimulus_in(31 downto 16);   -- vector array and
                                                        -- assigns the values
```

```
-- Put the output side (expected values) into a variable:
    S_expected := stimulus_in(15 downto 0);

    wait until falling_edge(Clk);              -- Wait until the clock goes
                                               -- back to '0' (midway through
                                               -- the clock cycle)

-- Check the expected value against the results in S:
    if (S /= S_expected) then
        err_cnt := err_cnt + 1;        -- Increment the error counter and
        assert false                   -- report an error if different
          report "Vector failure!" & lf &
          "Expected S to be " & vec2str(S_expected) & lf &
          "but its value was " & vec2str(S) & lf
          severity note;
    end if;
end loop;                             -- Continue looping through the file

    done <= '1';                      -- Set a flag when we are finished; this
                                      -- will stop the clock.

    wait;                             -- Suspend the simulation

end process;

end stimulus;
```

This test bench reads files of text "dynamically" during simulation, so the test bench does not have to be recompiled when test stimulus is added or modified. This is a big advantage for very large designs.

Using file I/O for test data can reduce the time required to add or modify test data.

What does the test vector file that this test bench reads look like? The following file (**testfib.vec**) describes one possible sequence of tests that could be performed using this test bench:

```
1000000000000000000000000000000000
0000000000000000000000000000000001
0000000000000000000000000000000001
0000000000000000000000000000000010
0000000000000000000000000000000011
0000000000000000000000000000000101
0000000000000000000000000000001000
```

```
0000000000000000000000000000001101
0000000000000000000000000000010101
0000000000000000000000000000100010
0000000000000000000000000000110111
0000000000000000000000000001011001
0000000000000000000000000010010000
0000000000000000000000000011101001
0000000000000000000000000101111001
0000000000000000000000001001100010
0000000000000000000000001111011011
0000000000000000000000011000111101
0000000000000000000000101000011000
0000000000000000000001000001010101
0000000000000000000001101001101101
0000000000000000000010101011000010
0000000000000000000100010100101111
0000000000000000000110111111110001
0000000000000000001011010100100000
0000000000000000010010100010001
0000000000000000000000000000000001
0000000000000000000000000000000001
0000000000000000000000000000000010
0000000000000000000000000000000011
0000000000000000000000000000000101
0000000000000000000000000000001000
```

Test data stored in files can be modified with any text editor, and no recompile is required when the data are changed.

This file could have been entered manually, using a text editor. Alternatively, it could have been generated from some other software package or from a program written in C, Basic or any other language. Reading text from files opens many new possibilities for testing and for creating interfaces between different design tools.

Although test vectors are quite useful for tabular test data, they are not particularly readable. In the last example of this chapter, we will describe how you can read and process test stimulus files that are more command-oriented, rather than simply being tables of binary values.

VHDL 1076-1993 File I/O

Before leaving the above example, let's see how it might look when coded using the 1076-1993 language features. In this version of the same test bench, we have modified the VHDL source file to reflect changes in the 1076-1993 standard. In addition to adding various syntax enhancements (such as allowing an **is** keyword to be used in a component declaration) for readability and consistency, the 1993 specification adds additional features for better control over the opening and closing of files. In the following source file, we use the **file_open** built-in function to open the test vector file.

```
----------------------------------------------------------
-- Test bench, VHDL '93 style
--
library ieee;
use ieee.std_logic_1164.all;
use std.textio.all;
use work.fib;    -- Get the design out of library 'work'

entity testfib is
end entity testfib;

architecture stimulus of testfib is
   component fib is
     port (Clk,Clr: in std_logic;
         Load: in std_ulogic;
         Data_in: in std_ulogic_vector(15 downto 0);
         S: out std_ulogic_vector(15 downto 0));
   end component fib;

   function str_to_stdvec(inp: string) return std_ulogic_vector is
     variable temp: std_ulogic_vector(inp'range) := (others => 'X');
   begin
     for i in inp'range loop
       if (inp(i) = '1') then
         temp(i) := '1';
       elsif (inp(i) = '0') then
         temp(i) := '0';
       end if;
     end loop;
     return temp;
   end function str_to_stdvec;

   function stdvec_to_str(inp: std_ulogic_vector) return string is
     variable temp: string(inp'left+1 downto 1) := (others => 'X');
   begin
     for i in inp'reverse_range loop
       if (inp(i) = '1') then
```

```vhdl
                    temp(i+1) := '1';
                elsif (inp(i) = '0') then
                    temp(i+1) := '0';
                end if;
            end loop;
            return temp;
        end function stdvec_to_str;

    signal Clk,Clr: std_ulogic;
    signal Load: std_ulogic;
    signal Data_in: std_ulogic_vector(15 downto 0);
    signal S: std_ulogic_vector(15 downto 0);
    signal done: std_ulogic := '0';

    constant PERIOD: time := 50 ns;

begin
    UUT: fib port map(Clk=>Clk,Clr=>Clr,Load=>Load,
                Data_in=>Data_in,S=>S);

    Clock: process
        variable c: std_ulogic := '0';
    begin
        while (done = '0') loop
            wait for PERIOD/2;
            c := not c;
            Clk <= c;
        end loop;
    end process Clock;

    Read_input: process
        file vector_file: text;

        variable stimulus_in: std_ulogic_vector(33 downto 0);
        variable S_expected: std_ulogic_vector(15 downto 0);
        variable str_stimulus_in: string(34 downto 1);
        variable err_cnt: integer := 0;
        variable file_line: line;

    begin

        file_open(vector_file,"tfib93.vec",READ_MODE);

        wait until rising_edge(Clk);

        while not endfile(vector_file) loop
```

```
      readline (vector_file,file_line);
      read (file_line,str_stimulus_in) ;
      assert (false)
         report "Vector: " & str_stimulus_in
         severity note;
      stimulus_in := str_to_stdvec (str_stimulus_in);

      wait for 1 ns;

      --Get input side of vector...
      Clr <= stimulus_in(33);
      Load <= stimulus_in(32);
      Data_in <= stimulus_in(31 downto 16);

      --Put output side (expected values) into a variable...
      S_expected := stimulus_in(15 downto 0);

      wait until falling_edge(Clk);

      -- Check the expected value against the results...
      if (S /= S_expected) then
         err_cnt := err_cnt + 1;
         assert false
            report "Vector failure!" & lf &
            "Expected S to be  " & stdvec_to_str(S_expected) & lf &
            "but its value was " & stdvec_to_str(S) & lf
            severity note;
      end if;
   end loop;

   file_close(vector_file);

   done <= '1';

   if (err_cnt = 0) then
      assert false
         report "No errors." & lf & lf
         severity note;
   else
      assert false
         report "There were errors in the test." & lf
         severity note;
   end if;
   wait;

end process Read_input;
```

```
end architecture stimulus;

-- Add a configation statement. This statement actually states the
-- default configuration, and so it is optional.
configuration build1 of testfib is
   for stimulus
      for DUT: fib use entity work.fib(behavior)
         port map(Clk=>Clk,Clr=>Clr,Load=>Load,
            Data_in=>Data_in,S=>S);
      end for;
   end for;
end configuration build1;
```

Reading Non-tabular Data from Files

You can use VHDL's text I/O features to read and write many different built-in data types, including such data types as characters, strings, and integers. This is a powerful feature of the language that you will make great use of as you become proficient with the language.

The standard text I/O features do not include functions for reading and writing standard logic data types.

VHDL's text I/O features are somewhat limited, however, when it comes to reading data that is not expressed as one of the built-in types defined in Standard 1076. The primary example of this is when you wish to read or write standard logic data types. In the previous example (the Fibonacci sequence generator), we made use of type coversion functions to read standard logic input data as characters. This method works fine, but it is somewhat clumsy. A better way to approached this common problem is to develop a reusable package of functions for reading and writing standard logic data. Writing a comprehensive package of such functions is not a trivial task. It would probably require a few days of coding and debugging.

Fortunately, one such package already exists and is in widespread use. This package, **std_logic_textio**, was originally developed by Synopsys. Synopsys allows the package to be used and distributed without restriction. We will use the

std_logic_textio package to demonstrate how you might read data fields from a file and write other data to another file (or, in this case, to the console or simulator transcript window).

The circuit that we will be testing with our test bench is a 32-bit adder-subtractor unit, the complete source code for which is provided on the companion CD-ROM. The test bench that we wish to write will read information from a file in the form of hexadecimal numeric values. The data file, which we will name TST_ADD.DAT, will include both the inputs and the expected outputs for the circuit. A listing of TST_ADD.DAT, containing a small number of test lines, is shown below:

```
0 00000001 00000001 00000002 0
0 00000002 00000002 00000004 0
0 00000004 00000004 00000008 0
0 FFFFFFFF FFFFFFFF FFFFFFFE 1
0 0000AAAA AAAA0000 AAAAAAAA 0
0 158D7129 E4C28B56 FA4FFC7F 0
1 00000001 00000001 00000000 0
1 A4F67B92 00000001 5B09846F 0
1 FFFFFFFF FFFFFFFE FFFFFFFF 0
1 FFFFFFFE FFFFFFFF 00000001 1
1 00000002 00000004 00000002 1
```

Special type conversion functions are required to read and write data in hexadecimal format.

The standard text I/O features defined in VHDL standard 1076 do not include procedures to read data in hexadecimal format, so we will make use of the **hread** procedure provided in the Synopsys **std_logic_textio** package. **Hread** accepts the same arguments as the standard **read** procedure, but allows values to be expressed in hexadecimal format. We will use **hread** to read the second, third and fourth fields in each line of the file, as these fields are represented in hexadecimal format.

Because the first and last fields of the data file are single-bit values of type **std_ulogic**, we will also make use of an overloaded **read** procedure provided in **std_logic_textio**. VHDL's built-in **read** procedure is not capable of reading **std_ulogic** values, so the **std_logic_textio** package includes additional **read** procedure definitions that extend **read** for these values.

Finally, we wish to display the results of simulation in the simulator's transcript window, so we use the **hwrite** and overloaded **write** procedures provided in **std_logic_textio** to format and display the data values. Once again, these are procedures that are not provided in the standard VHDL text I/O package.

```
library ieee;
use  ieee.std_logic_1164.all;
use work.all;
use std.textio.all;

library textutil;      -- Synposys Text I/O package
use textutil.std_logic_textio.all;

entity tst_add is
end tst_add;

architecture readhex of tst_add is
    component adder32 is
        port (cin: in std_ulogic;
            a,b: in std_ulogic_vector(31 downto 0);
            sum: out std_ulogic_vector(31 downto 0);
            cout: out std_ulogic);
    end component;
    for all: adder32 use entity work.adder32(structural);
    signal Clk: std_ulogic;
    signal x, y: std_ulogic_vector(31 downto 0);
    signal sum: std_ulogic_vector(31 downto 0);
    signal cin, cout: std_ulogic;

    constant PERIOD: time := 200 ns;

begin
    UUT: adder32 port map (cin, x, y, sum, cout);

    readcmd: process

        -- This process loops through a file and reads one line
        -- at a time, parsing the line to get the values and
        -- expected result.
        --
        -- The file format is CI A B SUM CO, with A, B and SUM
        -- expressed as hexadecimal values.
```

*The **std_logic_textio** package provided by Synopsys helps when reading standard logic data values from files.*

```
file cmdfile: TEXT;      -- Define the file 'handle'
variable line_in,line_out: Line; -- Line buffers
variable good: boolean;   -- Status of the read operations

variable CI, CO: std_ulogic;
variable A,B: std_ulogic_vector(31 downto 0);
variable S: std_ulogic_vector(31 downto 0);
constant TEST_PASSED: string := "Test passed:";
constant TEST_FAILED: string := "Test FAILED:";

-- Use a procedure to generate one clock cycle...
procedure cycle (n: in integer) is
begin
   for i in 1 to n loop
      Clk <= '0';
      wait for PERIOD / 2;
      Clk <= '1';
      wait for PERIOD / 2;
   end loop;
end cycle;

begin

-- Open the command file...

FILE_OPEN(cmdfile,"TST_ADD.DAT",READ_MODE);

loop

   if endfile(cmdfile) then  -- Check EOF
      assert false
         report "End of file encountered; exiting."
         severity NOTE;
      exit;
   end if;

   readline(cmdfile,line_in);    -- Read a line from the file
   next when line_in'length = 0;  -- Skip empty lines

   read(line_in,CI,good);     -- Read the CI input
   assert good
      report "Text I/O read error"
      severity ERROR;

   hread(line_in,A,good);     -- Read the A argument as hex value
   assert good
```

```
                              report "Text I/O read error"
                              severity ERROR;

                         hread(line_in,B,good);    -- Read the B argument
                         assert good
                            report "Text I/O read error"
                            severity ERROR;

                         hread(line_in,S,good);    -- Read the Sum expected resulted
                         assert good
                            report "Text I/O read error"
                            severity ERROR;

                            read(line_in,CO,good);    -- Read the CO expected resulted
                         assert good
                            report "Text I/O read error"
                            severity ERROR;

                      cin <= CI;
                      x <= A;
                      y <= B;

                      wait for PERIOD;   -- Give the circuit time to stabilize

                      if (sum = S) then
                         write(line_out,TEST_PASSED);
                      else
                         write(line_out,TEST_FAILED);
                      end if;
                      write(line_out,CI,RIGHT,2);
                      hwrite(line_out,A,RIGHT,9);
                      hwrite(line_out,B,RIGHT,9);
                      hwrite(line_out,sum,RIGHT,9);
                      write(line_out,cout,RIGHT,2);
                      writeline(OUTPUT,line_out);    -- write the message

                   end loop;

                wait;

             end process;

          end architecture readhex;
```

*Overloaded **read** and **hread** procedures accept arguments of type **std_logic_vector**.*

Creating a Test Language

As you can see, VHDL is a robust programming language with features that go well beyond what is needed to describe logic. In this, our final example, we will show how you can use the features of VHDL to describe a more complex test bench that reads non-tabular, command-oriented information from a text file, parses that information to determine the appropriate test inputs, and performs output value checking as specified in the file. This example will make use of a number of built-in functions for file and text manipulation, as well as text I/O functions for the parsing and display of a variety of different data types, including strings, characters and numeric values.

VHDL is a robust programming language with features that go well beyond what is needed to describe logic...

We'll use the language features of VHDL 1076-1993 for this test bench example, although the features of VHDL 1076-1987 could also be used to perform the same tasks, with some minor modifications.

The design we will be testing is a driving game that was inspired by the "ChipTrip" example first described by Altera Corporation using their AHDL PLD language. In our version of the design (which is described in more detail in *Electronic Design Automation for Windows: a User's Guide*, published in 1995 by Prentice Hall), the objective is to create a sequence of test inputs that will cause an imaginary work-weary engineer to proceed from his office to the beach, as quickly as possible, without getting a speeding ticket. To make the trip more interesting, our hero must stop and pick up a pizza on the way. The map of Figure 9-2 illustrates the possible routes that can be taken.

This map shows three different types of roads: freeways, commercial streets, and residential roads. The car being driven has only two possible speeds, fast and slow. When the car is driven slowly, it advances from one point on the map (say, from **Ramp1** to **Ramp2**) in a given period of time. When driven fast, the car proceeds twice as far. There is no speed limit on the freeway, so the car can travel at full speed without fear of getting a ticket. On commercial streets, the car may

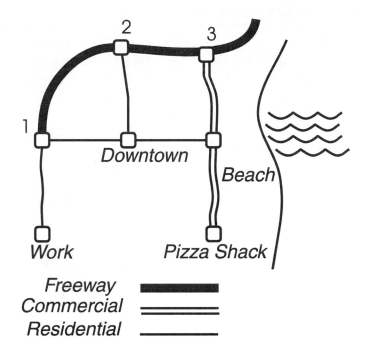

Figure 9-2: The driving game (inspired from Altera's ChipTrip design example) simulates a drive across town. The goal: get from the office to the beach (and pick up a pizza on the way) in the shortest amount of time without getting a ticket. Surf's up!

exceed the speed limit just once and get away with it. On residential roads, any attempt to drive fast will result in a ticket.

In our simulation, and in the underlying design description, a fixed period of time is represented by a single clock cycle. Inputs for the speed and initial direction of travel are represented by signals **Speed** and **Dir**. The location of the car at any point is represented internally to the circuit by a state machine, but it is kept hidden at the top level of the design and in the test bench itself. The current status and success or failure of a trip are observed on the signals **DriveTime**, **Tickets**, and **Party**, which tell the player how long the drive has taken, how many traffic tickets have accrued, and whether he or she has yet arrived at the beach with the pizza. The block diagram of Figure 9-3 shows the general layout of the elements of this design, which consists of two state machines, two counters and some additional logic. (The VHDL source files for the entire design are listed in Appendix F.)

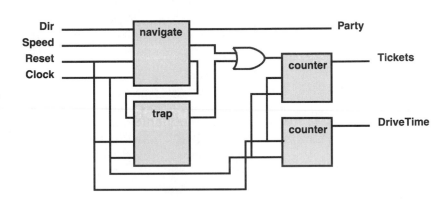

Figure 9-3: The driving game is composed of two state machines, a counter (repeated twice) and some random logic.

Test Bench Requirements

The test bench that we have written for this design reads symbolic test commands from a file, allowing the game to be easily tested and various driving scenarios to be described. There are three basic commands allowed in the test input file: **RESET**, **DRIVE** and **CHECK**.

The **RESET** command causes the game to be reset to its initial state. When this command is encountered in the input file, the **DriveTime** and **Tickets** counters are reset to zero, and the state machine that controls the drive is internally reset to state **Office**.

The **DRIVE** command has two additional arguments for direction and speed. It also specifies where the car should go, and how far, in one clock cycle. When **DRIVE** is encountered in the input file, the direction and speed arguments are parsed, the appropriate inputs (**Dir** and **Speed**) are assigned, and a single clock pulse is generated to advance the car to the next symbolic location on the map. In the test input file, the

257

direction is represented by the values **NORTH, SOUTH, EAST** and **WEST**, while the speed is represented by the values **FAST** and **SLOW**.

The **CHECK** command allows the value of signal **Party** or signal **Tickets** to be checked at any time to verify that the drive has proceeded as expected, and to allow the design to be tested for various sequences of directions and speeds.

Note:

To simplify the text field parsing in this example, we have intentionally made all commands read from the file the same size, in terms of the number of characters read for a given keyword. This includes the keywords "EAST " and "WEST ", "FAST " and "SLOW ", which in this implementation have an extra space appended onto them. To read string data (keywords) of random size, we would have to write a more complex token parser that reads the input line one character at a time, looking forward one character to identify delimiters such as spaces, tabs, and so on. This is not a difficult thing to do in VHDL, but it would clutter up an otherwise concise example.

The following sequence of test commands, entered into a file named **Tstpizza.cmd,** describes a number of possible test sequences for this game. Notice that the test file includes comments fields which are ignored by the test bench when the file is read:

The command language we have defined allows comment lines, and uses English-like keywords to describe the test sequence.

```
-- Test command file for getpizza example.
-- First try a winning game...
--
RESET
DRIVE NORTH SLOW        -- Cruise from work to Ramp 1
DRIVE NORTH FAST        -- No speed limit here!
DRIVE SOUTH FAST        -- Speed along the beach (1 warning)
DRIVE NORTH SLOW        -- Got the pizza; back to the beach!
CHECK PARTY 1
CHECK TICKS 0
--
-- Try getting some tickets this time...
--
RESET
DRIVE NORTH FAST        -- Speed through the residential area
DRIVE SOUTH SLOW        -- Cruise south to downtown
DRIVE EAST SLOW         -- Cruise to the beach
```

DRIVE WEST FAST -- Speed back through town
DRIVE EAST FAST -- Speed back to the beach
DRIVE SOUTH SLOW -- Go get the pizza
DRIVE NORTH SLOW -- Party time!
CHECK PARTY 1
CHECK TICKS 3

Test Bench Source File

The following source file, with explanatory comments, reads in the test file as described above and applies the necessary input stimuli to make the game proceed. **Assert** statements are used to verify that the test commands are properly entered, and to display errors when the **CHECK** command indicates a value for **Tickets** or **Party** other than those calculated during simulation.

```
--------------------------------------------------------
-- This test bench tests the getpizza driving game. It is
-- written with 1076-1993 features, and it makes use of
-- IEEE 1076.3 and the standard text I/O packages.
--
-- The test bench reads commands from a file (tstpizza.cmd)
-- to determine how the game should be "played".
--

use std.textio.all;

library ieee;
use  ieee.std_logic_1164.all;
use  ieee.numeric_std.all;    -- This design uses 'unsigned'

use work.game_types.all;   -- Contains types and constants

use work.pizzatop;

entity testgame is
end entity testgame;

architecture parser of testgame is
   component pizzatop is
      port(Clk,Reset: in std_logic;
          Speed: in tSpeed;
          Dir: in tDirection;
          Party: out std_logic;
```

```
                        Tickets: out std_logic_vector(3 downto 0);
                        DriveTime: out std_logic_vector(3 downto 0));
                   end component pizzatop;

                   signal Reset,Clk: std_logic;
                   signal Dir: tDirection;     -- std_logic_vector(1 to 2)
                   signal Speed: tSpeed;    -- std_logic
                   signal Party: std_logic;
                   signal DriveTime, Tickets: std_logic_vector(3 downto 0);

                   constant PERIOD: time := 40 ns;

              begin

                   UUT: pizzatop port map (Clk,Reset,Speed,Dir,Party,Tickets,DriveTime);

                   readcmd: process

                      -- This process loops through a file and reads one line
                      -- at a time, parsing the line to get the commands.

                      file cmdfile: TEXT;      -- Define the file 'handle'
                      variable L: Line;        -- Define the line buffer

                      variable keyword: string (1 to 5);  -- Used to get a keyword
                      variable c: character;    -- Used to read a single character
                      variable value: integer;  -- Used to read a numeric value
                      variable good: boolean;   -- For optional text I/O error checks

                      -- Use a procedure to generate one clock cycle...
                      procedure cycle (n: in integer) is
                      begin
                        for i in 1 to n loop
                          Clk <= '0';
                          wait for PERIOD / 2;
                          Clk <= '1';
                          wait for PERIOD / 2;
                        end loop;
                      end cycle;

                   begin

                      -- Open the command file...

                      file_open(cmdfile,"tstpizza.cmd",READ_MODE);
```

```
LOOP1: loop

    if endfile(cmdfile) then  -- Check EOF
        assert false
            report "End of file encountered; exiting."
            severity NOTE;
        exit LOOP1;
    end if;

    readline(cmdfile,L);         -- Read the line
    next LOOP1 when L'length = 0;  -- Skip empty lines

    read(L,keyword,good);        -- Read the command keyword
    assert good
        report "Text I/O read error"
        severity ERROR;
    case keyword is             -- Parse the command...
        when "RESET" =>
            Reset <= '1';
            cycle(1);
            Reset <= '0';
        when "DRIVE" =>
            read(L,c);        -- Eat the white space
            read(L,keyword);   -- Read the direction
            case keyword is
                when "NORTH" =>
                    Dir <= North;
                when "SOUTH" =>
                    Dir <= South;
                when "EAST " =>  -- Note the extra space
                    Dir <= East;
                when "WEST " =>
                    Dir <= West;
                when others =>
                    assert false
                        report "Unknown direction"
                        severity ERROR;
            end case;
            read(L,c);        -- Eat the white space
            read(L,keyword);   -- Read the speed
            case keyword is
                when "SLOW " =>
                    Speed <= SLOW;
                when "FAST " =>
                    Speed <= FAST;
                when others =>
```

```
                              assert false
                                  report "Unknown speed"
                                  severity ERROR;
                        end case;
                        cycle(1);
                   when "CHECK" =>
                      read(L,c);         -- Eat the white space
                      read(L,keyword);   -- Read the signal word
                      case keyword is
                         when "PARTY" =>
                            read(L,value);   -- Get a value
                            if value = 0 then
                               assert (Party = '0')
                                  report "Check failed on Party!"
                                  severity ERROR;
                            else
                               assert (Party = '1')
                                  report "Check failed on Party!"
                                  severity ERROR;
                            end if;
                         when "TICKS" =>
                            read(L,value);   -- Get a value
                            assert (UNSIGNED(Tickets) = value)
                               report "Check failed on Ticket count!"
                               severity ERROR;
                         when others =>
                            null;
                      end case;
                   when others =>    -- Only comments are valid here...
                      assert keyword(1 to 2) = "--"  -- Comment?
                         report "Unknown keyword"
                         severity ERROR;
               end case;

         end loop LOOP1;

      wait;

   end process;

end architecture parser;
```

You're On Your Way...

Congratulations! You've covered a lot of ground! Now it's time to start writing VHDL code of your own. You can use the code examples in this book, as well as additional examples found on the companion CD-ROM, as a place to start. The CD-ROM includes a full-featured, but design-size limited version of a commercial VHDL simulator, as well as other tools that will help you learn the language. Plus, you'll find helpful information gathered from a number of sources in the Appendices starting on the next page. We hope you've enjoyed our journey together and that you'll continue honing your skills in the exciting world of VHDL.

Appendix A: Getting The Most Out Of Synthesis

Bob Flatt, Metamor, Inc.

Today's synthesis tools offer engineers powerful computer-based methods for design entry and automated circuit creation. VHDL has emerged as an important standard for synthesis, and its language features offer great flexibility in how circuits can be described for synthesis.

But this flexibility in design description comes at a price: to make effective use of synthesis tools, it is crucial that you understand what the tools can and cannot do. It is also important to understand that synthesis tools, no matter how advanced, cannot design your logic for you. While synthesis tools do a great job of handling many design details, and can dramatically increase your productivity, they do have some inherent limitations. As a result, you must describe your logic designs with synthesis in mind. This article is intended to point out some common mistakes made by synthesis users, with the goal of making you a happier and more productive synthesis user.

Know Your Hardware

A common misconception is that a synthesis compiler "synthesizes VHDL." This is incorrect; the tool synthesizes *your design*, which is *expressed* in VHDL.

When using synthesis tools, the single most productive thing you can do is be aware of what kind of (and how much) hardware you are describing using your chosen HDL. Writing design descriptions without considering the hardware, and then expecting the synthesis tool to "do the design" for you, is a recipe for disaster. A common mistake is to create a design description, validate that description with a simulator, and assume that this "correct" specification must also be a "good" specification.

Understanding the hardware that you are specifying is the simplest rule for success. This is particularly important if you want to achieve critical timing goals. Conversely, the easiest way to fail at synthesis is to write a simulatable model of your design and then wonder why the synthesis tool didn't figure out all the details of implementation for you.

So what does synthesize mean in this context? Synthesis is simply a process of transforming a design specification into an implementation—in effect, converting a somewhat abstract design description into one that is more structured. This process is really nothing you couldn't do yourself; a synthesis tool simply handles the details of this somewhat complex transformation for you.

Common Synthesis Mistakes

To get the best results out of synthesis, and to achieve the highest possible portability between synthesis tools, it is important to describe your designs using well-established synthesis conventions, and to be aware of some common mistakes of first-time (and even more experienced) synthesis users. This section includes a few examples of typical circuit coding problems. These simple examples demonstrate real issues you will encounter when using synthesis tools. Some of these issues may seem obvious, while others may take you by surprise.

Testing for High-Impedance

The following example uses the IEEE 1164 standard logic data types and values in an attempt to describe the concept "if signal **sig** is floating." This is quite a reasonable test to perform in a simulation model. However, a synthesis tool has to transform this into a hardware element that matches this behavior.

```
if sig = 'Z' then  -- sig is std_logic
   -- do something
end if;
```

As written, the **if-then** test specifies a logic cell that looks at the drive of its fan-in, then outputs **true** if not driven, and **false** if driven **high** or **low**. Such a cell does not exist in most programmable silicon. (IEEE 1076.3 specifies that this comparison should always be false, so the statements inside the **if** are not executed, and no logic is generated.) Tests for 'Z' on inputs are therefore not meaningful in the context of synthesis.

Creating Long Signal Paths With Nested Ifs

Multiple nested **if** or **elsif** clauses can specify long signal paths. The following example describes a chain of dependent **if-then** statements:

```
if sig = "000" then
    -- first branch
elsif sig = "001" then
    -- second branch
elsif sig = "010" then
    -- third branch
elsif sig = "011" then
    -- fourth branch
elsif sig = "100" then
    -- fifth branch
else
    -- last branch
end if;
```

This code is an inefficient way to describe logic. A **case** statement would be much better. Why? Consider the test for the fourth branch, which depends on three previous tests and describes a long signal path, with a resulting potential logic delay.

```
case sig is
    when "000" =>
        -- first branch
    when "001" =>
        -- second branch
    when "010" =>
        -- third branch
    when "011" =>
        -- fourth branch
    when "100" =>
        -- fifth branch
    when others =>
        -- last branch
end case;
```

In practice, if the branches contain very little logic or if there are few branches, then there may be little difference in the amount of logic generated. However, the **case** statement generally results in a better implementation, so you should use **case** statements rather than **if-then-else** statements whenever possible.

Creating Long Signal Paths With Loops

Loops are very powerful, but each iteration of a loop replicates logic. A variable that is assigned in one iteration of a loop and used in the next iteration results in a long signal path. This signal path may not be obvious. An example where a long signal path is the expected behavior might be found in a carry chain (the variable **c** below):

```
function "+" (a,b: bit_vector) return bit_vector is
                              -- assumes a,b descending
   variable sum: bit_vector (a'length downto 0);
   variable c:bit := '0';
begin
   for i in a'reverse_range loop
      sum(i) := a(i) xor b(i) xor c;
      c := (a(i) and c) or (b(i) and c) or (a(i) and b(i));
   end loop;
   sum(a'length) := c;
   return sum;
end;
```

Examples of where this is not the expected behavior may be hidden in your code.

Synthesizing Code Optimized for Simulation

VHDL coders who have done extensive work creating simulation models know many "tricks of the trade" for creating simulation models that analyze and execute quickly. Such code, written for optimal simulation speed, will not usually be an optimal description of the logic for synthesis purposes.

In the following example it is assumed that only one control input will be active at a time. This description is efficient for simulation, but is a poor logic description for logic synthesis because the independence of the control signals is not described within the VHDL code:

```
out1 <= '0'; out2 <= '0'; out3 <= '0';
if in1 = '1' then
    out1 <= '1';
elsif in2 = '1' then
    out2 <= '1';
elsif in3 = '1' then
    out3 <= '1';
end if;
```

The independence of the control signals needs to be contained within the design description, or inefficient synthesis will result. The modified design description may be slightly slower during simulation, but will result in a smaller logic implementation after synthesis:

```
out1 <= '0'; out2 <= '0'; out3 <= '0';
if in1 = '1' then
    out1 <= '1';
end if;
if in2 = '1' then
    out2 <= '1';
end if;
if in3 = '1' then
    out3 <= '1';
end if;
```

Note that the issue here is not a long signal path, but an unclear specification of the design. The best optimizer in the world can't turn an inefficient algorithm into an efficient one. And an algorithm that is efficient from one viewpoint may not be efficient from another.

Using Port Mode Inout or Buffer: Which is Right?

Choosing whether to use mode **inout** or **buffer** for a port is most often determined by the requirements of the higher-level system interface, but you should understand the differences between these two modes in the event you have to choose between them.

Mode **inout** specifies a bi-directional dataflow. **Buffer**, like **out**, specifies unidirectional dataflow. There are very few occasions in hardware design when bi-directional is actually what you want, so you should use **buffer** for most cases in which you must (locally) read from an output port. New VHDL users often use **inout** when they have a logical output that they wish to read from, but mode **buffer** is actually more correct. Use **inout** only when you want to specify a signal path that is actually routed bidirectionally through a pin, such as when describing an I/O pad or (in a PLD) a pin feedback resource.

Synthesizing Type Conversion Functions

Type conversion functions generally specify no logic. They are most often intended to represent a simple buffer of zero delay that converts data of one type to data of another type. (This is not always the case, however.) Most such logic-free functions compile fairly quickly and, theoretically, result in no logic being generated. There is one common exception to this, however: a function that performs an array to integer conversion. Consider, for example, the following function:

```
function to_integer ( constant arg : bit_vector ) return natural is
    alias xarg: bit_vector(arg'length -1 downto 0) is arg;
                                    -- normalize direction with alias
        variable result: natural := 0;
        variable w: natural := 1;
    begin
        for i in xarg'reverse_range loop
            if xarg (i) = '1' then
                result := result + w;
            end if;
            if (i /= xarg'left) then
                w := w + w;
            end if;
        end loop;
        return result;
end to_integer;
```

This function will be slow to compile if **arg'length** is greater than around 16 to 24 bits (depending on your computer's speed/memory). This is the case because one of the "+" operators results in an adder being built for each iteration of the loop (even though the function describes no logic). These adders are presumably removed on data flow analysis, but this analysis can take an enormous amount of time to complete.

The solution to this problem is provided by synthesis vendors, who provide pre-analyzed ("approved") type conversion functions for array and numeric data types.

Depending On An Initial Value

The initial value of a signal or variable is the value specified in the object's declaration (if not specified, there is a default initial value). The initial value of such an object is its value when created. Signals and variables declared in processes are created at 'time zero'. Variables declared in subprograms are created when the subprogram is called.

The value at time zero has no clear meaning in the context of synthesis (what you get at system power-up is hardware-dependent). Therefore, the initial value of signals and process variables must be used with care. This issue does not arise with the initial value of variables declared in subprograms.

A good general rule is this: You should not depend on the initial value of signals or process variables if they are not completely specified in the process in which they are used. In this case, the compiler will ignore the time zero condition and use the driven value—effectively ignoring the single transition from the time zero state. If such signals or variables are not assigned, you may reliably use their initial value. Obviously, signals assigned in another process will never depend upon the initial value. For example:

```
signal res1: bit := '0';
begin
   process(tmpval,INIT)
   begin
     if (tmpval = 2**6 -1) then
        res1 <= '1';
     elsif (INIT ='1') then
        res1 <= '1';
     end if;
   end process;
```

In this case, **res1** is never assigned low—the code will be synthesized as a pull-up. However, during simulation at time zero, **res1** starts at '0', makes one transition to '1', then stays there. If this is really the intent, the proper solution is to use a flip-flop.

This design probably depends upon a wire floating low at power-up and probably has no realizable hardware implementation. A solution might be:

```
process(tmpval,INIT)
begin
   if (tmpval = 2**6 -1) then
      res1 <= '1';
   elsif (INIT ='1') then
      res1 <= '1';
   else res1 <= '0'; -- drive it low *****
   end if;
end process;
```

Making Assignments to Non-constant Array Indices

VHDL has many powerful features for manipulating arrays and slices of arrays. Synthesis tools differ in their ability to handle complex array and aggregate operations, but there are some general things to watch out for.

Consider the following assignment:

```
a(b) <= c;
```

If **b** is not a constant, then some care should be taken with this expression. This is because the statement means element **b** of array **a** gets the value of **c**. So far, so good. But it also means that all the other elements of the array get their previous value (they remain unchanged). In hardware this implies storage of data. If this assignment is not clocked, unwanted combinational feedback paths will likely be created.

A more typical usage might be:

```
a(b) <= c when rising_edge(clk);
```

If the assignment is clocked using the method shown above, the element select logic will drive the flip-flop clock enable control for an efficient implementation (assuming the synthesis tool is capable of generating clock enable logic).

Misunderstanding Don't-Cares

The semantics of the '-' element of **std_logic_1164** are not the same as the semantics of don't-cares as found in some PLD programming languages. The '-' in IEEE 1164 is a unique element of the 9-valued type **std_logic** and is not a wildcard when applied in a comparison operation.

For example, in the expression:

```
a <= "00010"
b <= a = "00---"
```

the signal **b** is never true!

If you wish to ignore comparison on some bits, then you must be explicit, as in:

```
b <= a(4 downto 3) = "00";
```

Creating Unintended Latches

Latches, whether intended or accidental, are inferred using incomplete specification in an **if** statement. The following example specifies a latch gated by **address_strobe**, which may not have been the intent.

```
process (address, address_strobe)
begin
  if address_strobe = '1' then
     decode_signal <= address = "101010";
  end if;
end process;
```

This description specifies that when **address_strobe** is '0', then the signal **decode_signal** holds its previous value, resulting in a latch implementation. In this case the intent is probably to ignore **decode_signal** when **address_strobe** is '0'. The following, more explicit description is more correct:

```
if address_strobe = '1' then
   decode_signal <= address = "101010";
else
   decode_signal <= false;
end if;
```

Note:

Synthesis tools will generally report to the user that latches have been inferred.

Creating Unintended Combinational Feedback

It is quite possible (and common) to accidentally specify combinational feedback paths by using variables (declared in a process) before they are assigned, or by incomplete specifications.

In the following example, if no element of the **ReadPtr(i)** array is ever equal to '1', **Qint** keeps its previous value. It may be a characteristic of the design that one bit of **ReadPtr** is always '1', but nothing in the specification says this is so. **Qint** is therefore incompletely specified and a feedback path exists, which includes **Qint** when **ReadPtr** is all zeros:

```
process(ReadPtr, Fifo)
begin
  for i in ReadPtr'range loop
    if ReadPtr(i) = '1' then
      Qint <= Fifo(i);
    end if;
  end loop;
end process;
```

We can code for this case by making certain **Qint** is always assigned. In this case, its value defaults to all zeros, and the unintended feedback path is removed:

```
process(ReadPtr, Fifo) begin
  Qint <= (others => '0'); -- because of possible comb feedback
  for i in ReadPtr'range loop
    if ReadPtr(i) = '1' then
      Qint <= Fifo(i);
    end if;
  end loop;
end process;
```

Conclusion

This article has helped illustrate some of the most common mistakes made by synthesis users and has highlighted the need for careful consideration of the intended hardware. Synthesis products such as those provided by Metamor can do a great job of converting your VHDL design descriptions into working hardware representations. Use synthesis tools wisely, and you will be a happier and more productive logic designer!

Appendix B: A VITAL Primer

Steven E. Schulz, Texas Instruments

VITAL (VHDL Initiative Toward ASIC Libraries) was recently adopted as Standard 1076.4 by the IEEE and is becoming increasingly popular as a way to express accurate timing information for high-perfomance ASICs and FPGAs. VITAL is a worldwide standards effort that enables signoff-quality, high-performance ASIC libraries in a VHDL environment that leverages from existing standards and ASIC cell practices.

This paper (which is adapted from a series of articles published in *Integrated Systems Design* magazine, formerly *ASIC&EDA*) will summarize the purpose and contents of VITAL, will explore some common misconceptions about VITAL, and will help to answer some of the most frequently asked questions about this important standard.

What is VITAL?

VITAL was formed to remove the barriers to creating VHDL ASIC libraries. As a result of this focus, VITAL has dramatically increased the availability of ASIC libraries across the industry and has achieved its original goal of sign-off simulation accuracy for ASICs. In the area of FPGAs, VITAL has been shown to be the ideal method with which to annotate post-route timing information to automatically-generated VHDL simulation models.

While the present scope is clearly focused on logic simulation for ASICs and FPGAs, follow-on and complementary efforts have leveraged from VITAL's foundation to address related issues in VHDL-based fault simulation (with at least one VITAL-based fault simulator now commercially available), timing verification, and synthesis.

VITAL's top two priorities have been clear from the start: 1) performance for complex, sub-micron ASICs, and 2) an aggressive development schedule. VITAL's developers consciously chose not to tackle extensive information modeling or theoretical ("what is the essence of timing?") considerations on the grounds that existing state-of-the-art techniques already in use can be, and need to be, brought into the VHDL environment quickly.

Simply put, VITAL consists of:

1. a set of timing procedures,

2. a set of primitives,

3. a mechanism for loading instance delays into simulation, and,

4. a specification document defining model development guidelines.

These four items are defined specifically with performance optimization in mind. The timing procedures include functions to select appropriate instance-specific delay values and perform common checks such as setup/hold, minimum pulsewidth, glitch detection, and so on. The primitives include the typical assortment of buffers, inverters, multiplexers, decoders, and latches, as well as truth and state table primitives. VITAL leverages the existing methodology for ASIC flows by allowing each ASIC foundry to maintain proprietary pre/post layout delay calculation tools, using Standard Delay Format (SDF) as the common file structure loaded into the simulator. The IEEE 1076.4 specification not only defines conventions for building ASIC libraries, but it also defines VHDL packages that implement the timing procedures and primitives, and it defines specific mapping between SDF constructs and VHDL.

Using VITAL, ASIC foundries are able to write cell libraries (models) based on standard primitives and timing functions. Instance-specific delays are linked to the models using generics with specified naming conventions. Simulators supporting VITAL then make use of an SDF loader to place the array of delay values into memory structures. The timing package (optimized into the kernel) can quickly access any desired value. The primitive package (also in the kernel) standardizes a set of primitives that may be used for the cells. Alternatively, a more behavioral coding style can be used, along with the same optimized timing, for megacells and cores (such as a DSP or 80486). The user synthesizes the design to a specific technology, and the structural VHDL references these library cells. SDF contains the detailed timing calculated for the entire circuit.

VITAL's primary goal is the specification and annotation of timing, but it has other applications and advantages as well. An example is VITAL's support of truth and state table primitives. Modeled after Verilog's "user-defined primitive" (UDP), these tables perform equivalent Mealy/Moore functionality, but with several added features requested by ASIC vendors. Whereas UDPs support multiple-input, single-output tables, VITAL's tables support multiple inputs and outputs. VITAL's tables also provide input and output state mapping, to minimize table size while satisfying accuracy requirements for various processes.

Another feature of VITAL is its support for negative hold checks. In today's sub-micron technologies, support for negative timing constraint checks has become increasingly important. Currently, negative hold checks require vendor-specific, non-standard methods when used in Verilog-based cells. Although VITAL cells will also require some special attention, negative hold checks are directly supported in the timing package with a standard modeling methodology, greatly easing the burden. Because negative constraints always require modeling with finer granularity signal delays, more effort is needed in any modeling environment.

And speaking of delays, VITAL also supports distributed delay models. While the path-delay style is generally preferred, this delay style is sometimes used by ASIC vendors to achieve maximum accuracy, at the expense of additional timing back-annotations and more sophisticated modeling requirements. VITAL treats distributed delays as a logical extension of hierarchy, where each sub-unit conforms to VITAL level-1 structures, almost as if the referencing circuit contained an added layer of hierarchy.

Who benefits from VITAL? First, ASIC designers enjoy improved portability of designs across platforms and a wider selection of libraries for most simulators. ASIC companies have benefited from reduced support costs after releasing portable, VITAL-compliant VHDL libraries. Because VITAL-compliant libraries are common across EDA platforms, ASIC vendors are able to reduce development and maintenance costs and reach more customers. EDA suppliers have benefited by focusing on tools instead of library issues, and they have been able to leverage from a standard set of optimization techniques.

Support From ASIC and EDA Industry

VITAL has always been fortunate to have strong corporate backing, with well over fifty companies signed up as VITAL members. The split between ASIC vendors, EDA vendors, and end-users is roughly 34%, 43%, and 22%, respectively.

VITAL has also been a focal point for other industry groups, such as the CAD Framework Initiative. CFI, as part of it's recent emphasis on ASIC library interoperability, has been working to identify potential opportunities for long-term convergence between VITAL and other related ASIC standardization efforts.

Cooperation between Verilog and VHDL/VITAL standards efforts has also been important in this effort. Open Verilog International's SDF working group has graciously offered to support enhancements to their Standard Delay Format, which will make VITAL's support for SDF "conditional delays" more elegant.

VHDL International has seen VITAL as among the most critical efforts for VHDL's continued growth, and it has been an active supporter in several ways. One benefit has been the VITAL ftp server (under **/vi/vital** at **vhdl.org**), where VITAL maintains the latest documentation and VHDL packages.

Of all industry groups, none is likely to have as great a future impact as the IEEE. The passage of VITAL as IEEE Standard 1076.4 was a culmination of years of effort on the part of many individuals. During this standarization process, I have witnessed an intense dedication of time and talent that brought VITAL to this point. Special thanks to Victor Berman, Bill Billowitch, Oz Levia, Victor Martin, Ray Ryan, Tom Senna, and Herman VanBeek, all of whom have been important contributors to the VITAL steering committee. It has been my honor and my pleasure to work with each of you.

Exposing Some Myths of VITAL

Now that we have explored the primary goals of VITAL, I would like to address a few common misconceptions surrounding VITAL. It has been quite gratifying to see so many hundreds of people learning about VITAL and developing products that use it. However, even more have only heard about VITAL through word of mouth and have formed erroneous conclusions based on misconceptions. I've addressed a few popular myths below, in hopes of replacing them with facts.

Myth: "If you are not an ASIC cell developer, VITAL won't impact you."

Fact: VITAL can provide key benefits for anyone who uses VHDL simulation:

- Accelerated primitives and timing check routines may be used in user-written models to gain nearly an order of magnitude speedup in VHDL simulation.

- VITAL-based libraries are designed to be portable across multiple toolsets, reducing the library availability problem, easing design reuse, and facilitating exchange with external design groups.

Myth: "VITAL doesn't support slope-dependent delays."

Fact: VITAL's architecture is flexible, relying on external delay calculation to supply all back-annotated data for the design simulation. If your ASIC supplier uses slope-dependent delays and outputs SDF, VITAL can support it too.

Myth: "VITAL's performance won't match gate-level simulators."

Fact: VITAL can be just as fast as any gate simulator for the very same technical reasons. Of course, VITAL doesn't determine performance; that's up to each implementation. Yet VITAL's architecture and modeling guidelines were designed to permit the same "tight" optimizations used in gate algorithms. VITAL acceleration is nothing more than placing constraints on the flexibility of VHDL within a defined scope. VHDL can simulate it all, but the portion recognized as VITAL-compliant can immediately execute subroutines as efficient as any other gate simulator's subroutines. Results are now available indicating that several released VHDL simulators out-perform Verilog-XL at several ASIC suppliers' sites.

Myth: "VITAL doesn't support minimum/typical/maximum delay modes."

Fact: Since any given simulation run only uses one of these sets of values, VHDL generics do not require duplication for each mode. All modes are supported in VITAL, based on the choice of SDF file used and/or the tool's mapping scheme from SDF.

Where To Learn More

If you are new to VITAL, I encourage you to take a closer look on VITAL's internet server, which may be accessed in several ways. Use ftp or gopher to vhdl.org under **/vi/vital**, or if you are on the Web, use the URL: **http://www.vhdl.org/vi/vital**. The IEEE 1076.4 working group uses this server to maintain all VHDL packages, IRs, example models, meeting minutes, and even a color slide tutorial describing more of the technical details of VITAL.

For those who have not been active in VITAL so far but are interested in its future development, it's not too late to get involved. You can add yourself to VITAL's email reflector by sending a message to **vital-request@vhdl.org**. To download the latest spec or VHDL packages, use the ftp server or use the automatic email-based file transfer feature. To learn more about becoming a VITAL member company, contact me directly at **ses@dadd.ti.com**.

Appendix C: Using VHDL Simulation

Introduction

Editor's note: The following tutorial was provided courtesy of Accolade Design Automation, Inc. A demonstration version of Accolade's PeakVHDL Personal Edition simulator can be found on the companion CD-ROM.

Accolade Design Automation's *PeakVHDL™* simulator is a design entry and simulation system that is intended to help you learn and use the VHDL language for advanced circuit design projects. The system includes a VHDL simulator, source code editor, hierarchy browser and on-line resources for VHDL users. You can use PeakVHDL to create and manage new or existing VHDL projects, and to run interactive simulations of those projects. Because VHDL is a standard language, you can use PeakVHDL in combination with other tools (including schematic editors, high-level design tools, and other tools available from third parties) to form a complete electronic design environment.

The Personal Edition PeakVHDL simulator described in this document, and supplied (in the form of a demonstration version) on the companion CD-ROM, allows VHDL design descriptions written in 1076-1987 standard VHDL (excluding its configuration features) to be entered and verified. If you wish, you can install this software from the CD-ROM and follow along with this tutorial. More advanced versions of PeakVHDL include support for 1076-1993 and are available from Accolade Design Automation.

Design Management

PeakVHDL includes many useful features that help you to create, modify and process your VHDL projects. The Hierarchy Browser, for example, shows you an up-to-date view of the structure of your design as you are entering it. This is particularly useful for projects that involve multiple VHDL source files (called *modules*) and/or multiple levels of hierarchy.

The Entity Wizard helps you create new VHDL design descriptions by asking you a series of questions about your design requirements and generating VHDL source file templates for you based on those requirements.

The built-in dependency ('make') features help you streamline the processing of your design for simulation and for synthesis. For example, when you are ready to simulate your design, you simply highlight the design unit (whether a source file module, entity, architecture, etc.) you wish to have processed and click a single button. There is no need to compile each VHDL source file in the design or keep track of your source file dependencies. PeakVHDL does it for you.

Simulation Features

PeakVHDL's built-in simulator is a complete system for the compilation and execution of VHDL design descriptions. The simulator includes a VHDL compiler, linker, and simulator. VHDL design descriptions are compiled into a 32-bit native Windows executable form. When run, these executable files interact with the PeakVHDL system to allow interactive debugging of your circuit. The PeakVHDL compiler, linker and simulator are also native 32-bit Windows applications, so it is recommended that you install the software using either Windows 95 (or later) or Windows NT. (It is possible to operate the sofware on Windows 3.1, using the OLE32s extensions provided. Refer to the on-line "readme" information for more details.)

Simulation results (in the form of graphical waveforms and/ or textual output) can be easily saved for later use (including detailed timing analysis using SynaptiCAD's *The Waveformer* software, which is also provided on the CD-ROM and discussed in Appendix D) or printed on any Windows-compatible printer.

Understanding Simulation

Simulation of a design description using PeakVHDL involves three major steps:

1. Compiling the VHDL modules into an intermediate object file format.

2. Linking the object files to create a simulation executable.

3. Loading the simulation executable and starting the simulation.

Each of these three steps is represented by an icon button in the PeakVHDL application. If the dependency features of the application are enabled, PeakVHDL will check the date and time stamps of files, and it will examine the hierarchy of your design to determine which files must be compiled and linked at each step.

When a simulation executable has been successfully linked and loaded, the Waveform Display appears and you are ready to start a simulation run. If you have used VHDL's assert or text I/O features, you can interact with the simulation using both text and waveform methods.

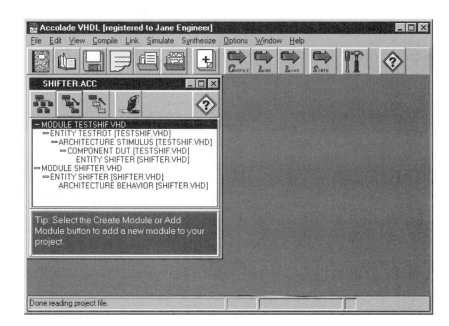

Figure C-1. With a project loaded into the Accolade VHDL application, the Hierarchy Browser becomes your primary view of the files and design units making up the project.

Loading the Sample Project

To help you understand how simulation works, we'll load and run a sample project supplied with the PeakVHDL simulator. This sample project (a barrel shifter) is located in the **Examples\Shifter** directory installed with your PeakVHDL software.

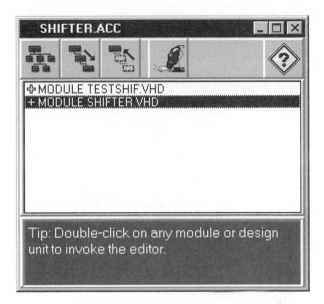

Figure C-2. Use the plus and minus icons to examine the hierarchy for each module.

To load the sample project, invoke the VHDL Simulator and select the **Open Project** item from the **File** menu. Navigate to the **Examples\Shifter** directory and choose the **Shifter.ACC** file. The Project will be loaded as shown.

Using the Hierarchy Browser

After you open the project, you will see that there are two modules shown in the Hierarchy Browser. These modules (TESTSHIF and SHIFTER) are VHDL modules that were entered to describe the operation of the sample circuit. TESTSHIF is a test bench for the circuit, while SHIFTER describes the function of the shifter circuit itself. You can examine or modify either of these modules by double-clicking on them to invoke a Source Code Editor window.

Note that the Hierarchy Browser does not provide any immediate indication of which module represents the "top" of your design. (The order of modules appearing in the Hierarchy Browser is not significant.) This is because you may choose to select different top-level modules depending on whether you are invoking simulation or synthesis, or whether you want to simulate just a portion of the circuit or simulate the entire circuit. You may also have more than one top-level test bench in your project.

You can, however, display the hierarchy and file dependencies for any module displayed in the Hierarchy Browser. By clicking on the small white "+" icons to the left of each module, you can view the hierarchy for that module.

The Hierarchy Browser is the point from which you initiate all processing of your design, from compiling and linking to synthesis and simulation.

Compiling Modules for Simulation

Setting Compile Options

Before compiling this design, take a moment to examine the compile options that have been selected. First, highlight the module SHIFTER in the Hierarchy Browser.

Next, select **Compile** from the **Options** menu (or click on the Options button) to bring up the Compile Options dialog. The options should be set as shown in Figure C-3.

The options set are:

- **Bottom up to selected**. This option tells the compiler to examine the dependencies of the project and compile lower-level VHDL modules before compiling higher-level modules that depend upon them.

- **Compile only if out of date**. This option enables the date and time stamp checking features so that modules are not compiled unless they are out of date. This can save time when you are compiling a large project repeatedly (such as when fixing syntax errors in higher-level modules).

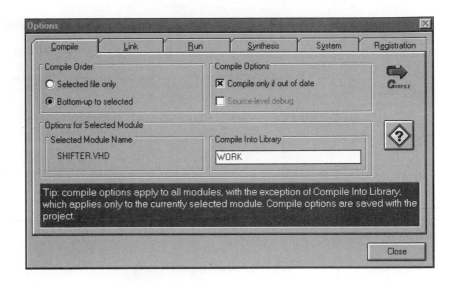

Figure C-3. The Options dialog allows you to set processing options for compiling, linking and simulation.

- **Compile Into Library**. This option specifies that the current module, SHIFTER, is to be compiled into a named library, in this case WORK.

When you have verified that the options are set to these values, select the Close button to close the Options dialog.

Starting a Compile

The next step is to compile the source modules. With the **Bottom up to selected** option selected, you have two options:

1. You can first compile the SHIFTER module, then compile the TESTSHIF module, or,

2. You can simply compile the TESTSHIF module and let the dependency features automatically compile the lower-level SHIFTER module.

Use the second method to compile the two source files. To start the compile, highlight the TESTSHIF module in the Hierarchy Browser and click the Compile button.

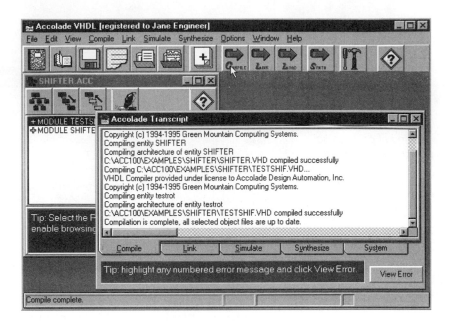

Figure C-4. During processing, status and other messages are displayed in a scrollable transcript window. If you wish, you can save the contents of the window or print it to any Windows compatible printer.

During compilation, status and error messages are written to the Transcript window. If you wish, you can save these messages to a file or print them directly.

When compiled, each VHDL source file (module) in the project is processed to create an intermediate output file (an *object file*). These files, which have a .O file name extension, must be linked together to form a *simulation executable*.

Note:
If system or other errors occur during the processing of this design example, you should check to make sure you have properly installed and registered the PeakVHDL software. The software will not operate without first being registered.

Windows 3.1 Users Note:
*The PeakVHDL software is a 32-bit application intended for use on Windows 95 or Windows NT. It will run on most Windows 3.1 systems. However, if you encounter problems running the software on Windows 3.1, we recommend you download the latest demo version of PeakVHDL from the Accolade Design Automation Web site at URL **www.acc-eda.com**, and/or use Windows 95 or Windows NT.*

Linking Modules for Simulation

Setting Link Options

Before linking the modules, take a moment to examine the link options. First, highlight the module TESTSHIF in the Hierarchy Browser as shown in Figure C-5.

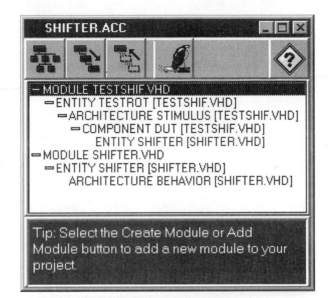

Figure C-5. To link the project for simulation, you must first select a top-level module, entity or architecture.

Next, select **Link** from the **Options** menu (or click on the options button and choose the Link tab) to bring up the Link Options dialog. The options should be set as shown in Figure C-6.

The options set are:

- **Update object files before linking**. This option tells the linker to examine the dependencies of the project and compile lower-level VHDL modules before linking. (If the **Compile only if out of date** option is specified in the Compile Options dialog, only those source files that are out of date will be recompiled.)

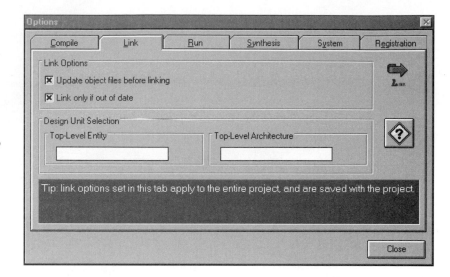

Figure C-6. The link options allow you to specify a default top-level entity and architecture. They also give you control over the automatic update features of the interface.

- **Link only if out of date**. This option enables the date and time stamp checking features so that the modules are not re-linked unless the simulation executable is out of date.

- **Design Unit Selected**. These fields allow you to specify a default top-level entity and architecture for the selected module. Because this project does not include multiple entities and architectures within the top-level module, you can leave these fields blank.

When you have verified that the options are set to the values shown, select the Close button to close the Options dialog.

Starting a Link Operation

To link a project, the linker collects all object files required for the selected top-level module and combines these object files with any libraries you have selected (such as the IEEE standard logic library) to create the simulation executable.

To link the design and create a simulation executable, highlight the TESTSHIF module in the Hierarchy Browser and click on the Link button.

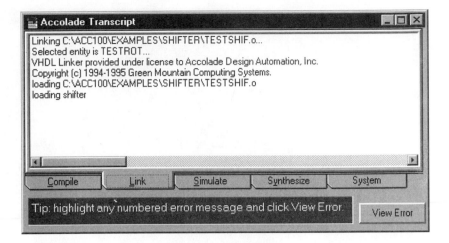

Figure C-7. During the link process, the object files (created as a result of compilation) are combined with any external libraries to create a simulation executable.

During the linking process, messages will be written to the transcript window.

The result of linking is a simulation executable file ready to be loaded for simulation.

Using the Accolade Simulator

Setting Simulation Options

Before loading the simulation executable, take a moment to examine the simulation options. First, make sure the TESTSHIF module is still highlighted in the Hierarchy Browser.

Next, select **Simulate** from the **Options** menu (or click on the options button and choose the **Simulate** tab) to bring up the Simulate Options dialog. The options should be set as shown in Figure C-8.

The options set are:

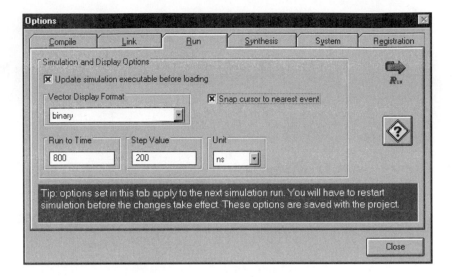

Figure C-8. *Before loading the simulation executable, it is a good idea to set the initial run time and step intervals.*

- **Update simulation executable before loading**. This option tells the simulator to examine the dependencies of the project and, if necessary, compile lower-level VHDL modules and re-link them before loading.

- **Vector Display Format**. This option specifies how vector (array) data types should be displayed. Select **binary** for this field.

- **Snap cursor to nearest event**. This option specifies that waveform cursors should 'snap' to an event when they are placed or moved in the waveform display.

- **Run to Time and Step Value**. These fields allow you to specify a default amount of time that the simulation should run. These values can be changed during simulation if necessary.

- **Unit**. This field specifies the unit of time to be used during simulation.

When you have verified that the options are set to the values shown, select the Close button to close the Options dialog.

Loading the Simulation Executable

To load the simulation executable, make sure the TESTSHIF module is selected in the Hierarchy Browser and click on the Load button. The simulation executable will be loaded and the waveform display window will appear as shown in Figure C-9.

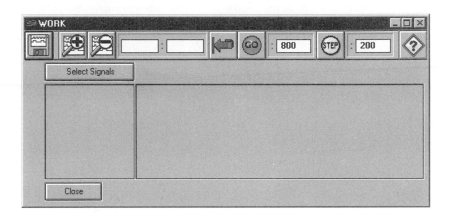

Figure C-9. When a simulation executable has been loaded, a blank Waveform Display window appears.

Selecting Signals to Display

Before you start the simulation, you must select one or more signals to observe. The Signal Selection dialog is shown in Figure C-10.

The Signal Selection dialog has two primary windows. The Available window lists all the signals available to be observed in the simulation, while the Displayed window shows a list of signals to be displayed. Use the **Add**, **Insert**, and **Remove** buttons to change the Displayed window contents, then click the **Display** button to add the Displayed signals to the Waveform Display.

Note:
The latest Displayed signals are saved with your project, so they are immediately available when you return to the project later.

297

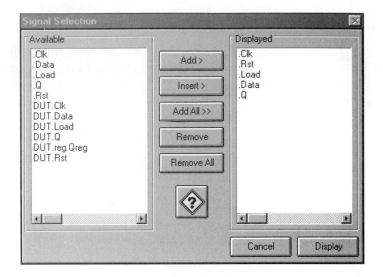

Figure C-10. Use the Signal Selection dialog to specify which signals you want to observe during simulation.

Starting a Simulation

After you have selected signals to observe during simulation, you can click on the **Go** button to start the simulation.

Simulation will run until either:

1. The specified simulation end time (duration) has been reached, or

2. All processes in your project have suspended.

Click on the **Go** button to simulate this project and generate a waveform. The resulting waveform should look something like the following (you can use the Zoom In button and the horizontal scroll bar to change the display range as shown):

Working with Waveforms and Cursors

The PeakVHDL Waveform Display has a variety of features for examining waveform results, saving waveforms to files and comparing simulation runs. For more information about these features, refer to the on-line help information.

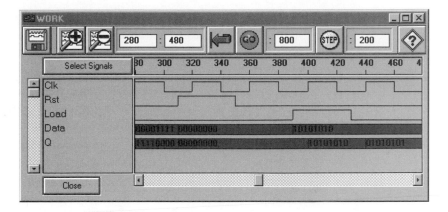

Figure C-11. After you have selected the GO button, the resulting waveform is displayed. Use the Zoom In and scrolling controls to view all or part of the waveform.

The Waveform Display also includes selectable cursors that can be used to accurately measure the distances between events and view the timing relationships between events on different signals.

To add a cursor to the Waveform Display, simply click the mouse button within the Waveform Display window. A new cursor is displayed each time you click the mouse button. By clicking and holding the mouse button, you can drag a new or existing cursor to any location on the Waveform Display. If the **Snap cursor to nearest event** option is selected, the selected cursor will "snap" to whatever event is nearest the location of the mouse pointer:

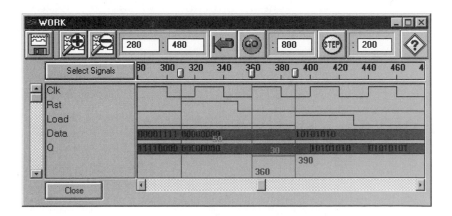

Figure C-12. Clicking once within the display will cause a measurement cursor to appear. You can add multiple cursors and drag them into position with the mouse.

299

To delete any cursor, click once on the small deletion button that appears in the ruler area of the Waveform Display.

Summary

This introductory tutorial has covered only the basics of VHDL simulation using the PeakVHDL Simulator. The simulator supports additional features such as text I/O that allow your VHDL design to interact with the simulation environment, and it includes features such as the Entity Wizard to help you create new design descriptions. A Professional Edition adds features such as source-level debugging and advanced editing and design management features to make finding and fixing errors in your designs easier and faster.

To learn more about design entry and simulation using VHDL, refer to the on-line help and Guided Tour information installed with PeakVHDL. Also, be sure to examine the VHDL examples provided with PeakVHDL. These examples demonstrate a variety of ways to use VHDL for describing and testing logic designs.

In addition to the examples and other information provided with PeakVHDL, you should take the time to visit the Accolade Design Automation home page at the following URL: **http://www.acc-eda.com**. This home page includes updated software, answers to frequently asked questions, and an on-line introduction to VHDL and VHDL design techniques that will help you get started with the language.

Appendix D: Test Bench Generation from Timing Diagrams

Donna Mitchell, SynaptiCAD, Inc.

Introduction

Editor's note: The following tutorial has been provided courtesy of SynaptiCAD, Inc. A demonstration version of SynaptiCAD's The Waveformer product is included on the CD-ROM and can be used to follow along with this tutorial.

Surveys of VHDL users have indicated that the generation of VHDL test benches typically consumes 35% of the entire front-end ASIC design cycle. It is clear that automation of test bench generation would significantly reduce the costs of VHDL-based design. This section describes a method for automating the generation of VHDL test benches using a combination of graphical and equation-based methods.

Background: What is a Test Bench?

After a VHDL model is written, it must be tested to verify correct operation. To do this, the designer provides a VHDL simulator with a set of input signal transitions (stimulus vectors) to exercise the circuit. These stimulus vectors are usually grouped together in an entity called a **test bench.** Grouping the stimulus vectors together in an isolated entity provides a means to reuse the stimulus vectors as the VHDL code for the circuit changes (e.g. as the simulation models are changed from behavioral models to structural models).

In VHDL there are two ways to write stimulus vectors: using **wait** statements or using **transport** statements. Transport-based test benches are smaller and easier to read than wait-based test benches, but wait-based test benches are easier to understand when single-stepping through a simulation to debug it. Figure D-1 shows examples of both wait-based and transport-based test benches.

Writing a VHDL test bench is one of the most tedious and error prone parts of the simulation process. In VHDL you are required to specify the time for each signal transition. It is easy to make mistakes when writing an HDL test bench, because it is difficult to visualize the temporal relationships between transitions on different waveforms. For instance, take a look at the **transport** statements in the listings of Figure D-1 and see if you can figure out if the first rising edge of **SIG0** comes before the rising edge of the third cycle of clock **clk**. This is where a graphical timing diagram editor like The WaveFormer excels. Figure D-2 shows a timing diagram that generated the code in Figure D-1. Looking at the timing diagram, it is easy to see that the **SIG0** transition does occur before the rising edge of the clock.

Transport VHDL Test Bench

```
library ieee, std; use std.textio.all;
use ieee.std_logic_1164.all;
entity testbench is
  port( CLK0: out std_logic;
    SIG0 : out std_logic;
    SIG1 : out integer;
    SIG2 : out std_logic);
end testbench;
architecture test of testbench is
begin
  process
  begin
    CLK0 <= '1'; wait for 0 ps;
    while true loop
      CLK0 <= '0'; wait for 25000 ps;
      CLK0 <= '1'; wait for 25000 ps;
    end loop;
  end process;
  process
  begin
    SIG0 <= transport '1' after 0 ps,
                '0' after 50000 ps,
                '1' after 90000 ps,
                '0' after 175000 ps;
    SIG1 <= transport 4 after 0 ps,
              16 after 55000 ps,
               9 after 120000 ps,
              12 after 180000 ps,
               3 after 235000 ps;
    SIG2 <= transport '0' after 0 ps,
                '1' after 20000 ps,
                '0' after 40000 ps,
                '1' after 60000 ps,
                '0' after 80000 ps,
                '1' after 100000 ps,
                '0' after 120000 ps,
                '1' after 140000 ps,
                '0' after 160000 ps,
                '1' after 170000 ps,
                '0' after 180000 ps,
                '1' after 190000 ps,
                '0' after 200000 ps,
                '1' after 210000 ps,
                '0' after 220000 ps,
                '1' after 230000 ps,
                '0' after 240000 ps;
      wait;
  end process;
end test;
```

Wait VHDL Test Bench

```
library ieee, std;
use std.textio.all;
use ieee.std_logic_1164.all;
entity testbench is
  port( CLK0: out std_logic;
    SIG0 : out std_logic;
    SIG1 : out integer;
    SIG2 : out std_logic);
end testbench;
architecture test of testbench is
begin
  process
  begin
    CLK0 <= '1'; wait for 0 ps;
    while true loop
      CLK0 <= '0'; wait for 25000 ps;
      CLK0 <= '1'; wait for 25000 ps;
    end loop;
  end process;
  process
  begin
    SIG0 <= '1';
    SIG1 <= 4;
    SIG2 <= '0';
    wait for 20000 ps;
    SIG2 <= '1';
    wait for 20000 ps;
    SIG2 <= '0';
    wait for 10000 ps;
    SIG0 <= '0';
    wait for 5000 ps;
    SIG1 <= 16;
    wait for 5000 ps;
    SIG2 <= '1';
    wait for 20000 ps;
    SIG2 <= '0';
    wait for 10000 ps;
    SIG0 <= '1';
    wait for 10000 ps;
    -- More wait statement code
    -- removed because of space limitations
    wait for 848 ps;
    wait;
  end process;
end test;
```

Figure D-1: *VHDL test benches using* **transport** *and* **wait** *statements.*

Automating Test Bench Generation

The WaveFormer is a timing diagram editor with the ability to generate and translate stimulus vectors to a variety of different simulation and documentation formats (including HDL test bench formats). Although specialized tools for translating stimulus vectors have existed for several years, these tools have traditionally been high-priced niche products that don't offer the sophisticated timing analysis and documentation features of timing diagram editors. Timing diagram editors like The WaveFormer allow users to interactively perform timing analysis on their waveforms and document timing relationships using delays, setups, holds, and textual information. These features, along with the ability to generate test benches and import results from simulations and test equipment, make The WaveFormer the ideal environment for waveform manipulation and analysis.

Tutorial: Generating Test Benches with The WaveFormer

In this tutorial we will demonstrate how to use The WaveFormer to generate VHDL test benches. You can perform the tutorial using The WaveFormer evaluation version located on the CD-ROM at the back of this book. If you are running Windows 3.1 then you will also have to install the Win32s libraries which are also located on the CD-ROM (Windows 95 and NT users do not need the Win32s libraries). A SunOS UNIX evaluation version can be downloaded from SynaptiCAD's web site at **http://www.syncad.com/**. Figure D-2 is the completed tutorial timing diagram.

Figure D-2: *The WaveFormer timing diagram that generated the code shown in Figure D-1.*

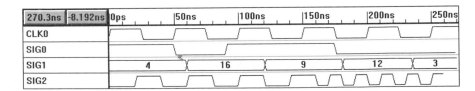

Figure D-3: *The WaveFormer timing diagram editor.*

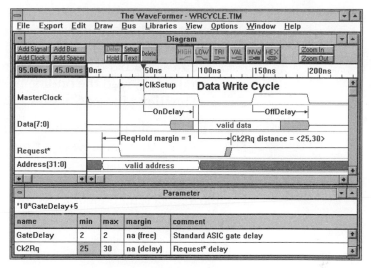

Automatically Generate a Clock Signal Using The WaveFormer

The WaveFormer supports automatic generation of periodic signals (e.g. a 66MHz clock with a 40% duty cycle). To create a periodic waveform, you enter information about the clock, such as frequency and duty cycle, and The WaveFormer generates an appropriate waveform. We begin the tutorial by creating a clock named **CLK0** with a period of 50ns (20MHz).

Add clock CLK0

1. Click on the **Add Clock** button (top left hand corner of the screen). This will cause the **Edit Clock Parameters** dialog box to appear.

2. Enter the name **CLK0** in the **Label** edit box. Enter **50** in the **Period** box. Make sure the **MHz/ns** radio button is selected. (The frequency will change to match the new period value when you move the selection to another edit box.)

3. Left mouse click on the OK button to close the dialog and create the clock.

Although we only used a value to set the period of this clock, clock attributes can also be based on formulas. Clock formulas can be defined in terms of timing parameters and the attributes of other clocks. For more information on clock parameters, read the on-line help **Table of Contents/ Design Functions Index/Chapter 2: Clocks** entry.

Your clock should look like the one diagrammed in Figure D-2. If you made a mistake entering the clock information, double left click on a clock segment (area between two edges) to reopen the **Edit Clock Parameters** dialog box. Double left clicking on a clock edge opens up the **Edge Properties** dialog box which displays the edge time.

Drawing signals in The WaveFormer

Next we will draw signals SIG0 and SIG1.

Add SIG0 and SIG1

1. Left mouse click on the **Add Signal** button two times to add two signals.

Sketch the waveform for SIG0 using high and low segments

1. Left click on the **LOW** button, then left click on the **HIGH** button. This sets up the current drawing and next drawing state (notice that the HIGH button is red and the LOW button has a small red T at the top to indicate the toggle state).

2. Next, put the mouse inside the drawing window at the same level as SIG0 and left click at approximately 50ns, 90ns, 175ns, and 270ns. Your waveform should look like SIG0 of Figure D-2.

Sketch the waveform for SIG1 using just valid segments

1. Left click on the **VAL** button to turn it red, and left click again on the **VAL** button to move the red T to the button. This causes the VAL button to stay selected when you draw waveform segments.

2. Put the mouse inside the drawing window at the same level as SIG1 and left click at approximately 55ns, 120ns, 180ns, 235ns, and 270ns. Your waveform should look like SIG1 of Figure D-2 without the state values (we will add the state values later in the tutorial).

If you made a mistake drawing the waveforms, use the following editing techniques to correct the waveform:

3. Drag and drop signal edges.

4. Select a segment, then click a new state button to change the state of the segment.

5. Select an edge, a segment, or a signal name and hit the delete key to delete the selected object.

6. Double click on an edge or signal name to edit it.

7. Double click on a segment to insert a new segment.

Generate Signals using Waveform Equations

If a waveform has a periodic or known pattern, then it is easier to specify the signal as an equation rather than drawing each repetitive piece of the signal. The WaveFormer has an equation interface for just those types of signals. For example, SIG2 is a periodic waveform with a change in frequency from 25Mhz to 50Mhz which is generated from the following temporal equation:

SIG2 (20ns=0 20ns=1)*4 (10ns=0 10ns=1)*5

Generate SIG2 using the above equation

1. Choose the **Export\Draw Waveform Equation** menu option, which will open a dialog of the same name.

2. Type the above equation into the edit box in the dialog, and choose the **OK** button. This creates a signal in The WaveFormer called SIG2 with an initial frequency of 25MHz (period = 20+20 = 40ns) for 4 cycles and switches to 50Mhz for 5 more cycles. This type of waveform is tedious to draw by hand, but it can be concisely expressed as a temporal equation.

If a signal named SIG2 had already existed when this equation was entered, the resulting waveform would have been added on to the end of the signal. This ability to concatenate waveforms to an existing signal, in combination with the scripting language of The WaveFormer (discussed further in the import/export section), makes it possible to create extremely complex waveforms.

Advanced data types: Extended State Info. and Object Properties

The WaveFormer supports both simple **std_logic** signals and complex user-defined data types for VHDL test benches. By default, all signals are assumed to have a type of **std_logic** and a direction of **out** (CLK0, SIG0, and SIG2 will use the defaults for this tutorial). To model complex data types and user defined states, The WaveFormer has *object properties* and *extended state* features which allow the user to attach arbitrary data to objects in the timing diagram. This means that The WaveFormer can support all VHDL data types, including arrays and user-defined enumerated types.

To demonstrate the *object properties* and *extended state* features, we will add information to SIG1 so that it is exported as a VHDL signal with a type of **integer**, and each state in SIG1 will have an integer value.

Change the type of SIG1 to integer.

1. First choose the **Export\Edit Object Properties** menu option which opens a dialog box of the same name.

2. From the **Object Type** drop-down listbox at the top of the dialog, select **Signal Properties**. Select the SIG1 from the **Object Name** listbox.

3. Type **VHDLtype** into the **Property** edit box. Type **integer** into the Value edit box.

4. Click the **ADD** button to add it to the signal. Click the **OK** button to close the dialog. (You could also change the direction by adding a **VHDLdir** property and value, but we have not done this for the sake of simplicity.)

Add the integer state values to SIG1.

1. First select a segment of SIG1 by left clicking on it (a selected segment has a dotted box drawn around it).

2. Next, left click on the **HEX** button; this will open the **Edit Bus** dialog.

3. Type an integer into the ExState edit box, and choose the next or previous buttons (**alt-n** or **alt-p** keys) to edit the segments on either side. (In Figure D-2, SIG 1 has 4, 16, 9, 12, and 3 signal state values.)

4. Click **OK** when you are done.

Note:

*The type and state values of SIG1 can be anything you want. For example, the **VHDLtype** could be MYCOLOR and extended state information could be RED, BLUE, BRIGHT BLUE, GREEN, and PURPLE.*

Export to VHDL

Next, we will export the timing diagram to a VHDL test bench.

To export the timing diagram to VHDL:

1. Choose the **Export\Export Signals As** menu option. This will open the Export Dialog.

2. Choose the type of script using the **Save File as Type** list box in the lower left corner of the Export Dialog Box. We used **VHDL transport(*.vhd)s** and **VHDL wait(*.vhd)s** to generate the test benches in Figure D-2.

3. Pick a file name and click the **OK** button. The WaveFormer will produce a file with the timing data in that format.

4. View the generated file using a text editor (e.g. Notepad).

The generated file should look approximately like the file displayed in Figure D-1. (Because of drawing differences, the times for signal transitions may differ slightly.) Congratulations! You have completed the tutorial and generated a VHDL test bench.

Customizing VHDL output using Waveperl Scripting Language

The WaveFormer uses a scripting language based on Perl (Practical Extraction and Report Language) to import and export waveform stimulus. The use of a scripting language means users can modify the output of existing translation scripts and even create their own scripts to support waveform formats used by custom test equipment and internally developed software.

Support for Simulation and Other Environments

The WaveFormer currently creates input stimulus for:

- Standard VHDL

- Standard Verilog

- SPICE (analog & digital formats supported)

- Viewlogic's Viewsim simulator

- Mentor's QuickSim simulator.

Waveform import is supported for:

- Accolade's PeakVHDL simulator

- WellSpring's Verilog simulator

- Viewlogic's Viewsim simulator

- Boulder Creek Engineering's Pod-A-Lyzer (a PC-based logic analyzer).

We frequently add new scripts, so send us an email at **syncad@bev.net** and tell us what import/export formats you would like us to add. Also, all the scripts that ship with The WaveFormer are editable by the user. This is particularly useful because most waveform timing information is written in only a couple of different formats. If you have a proprietary format, there is probably a Waveperl script that does most of what you need. Waveperl scripts are normal text files that can be edited with any text editor.

Beyond Simple Stimulus Generation

In this tutorial we have a demonstrated (1) how to create a simple timing diagram using drawn and equation-generated signals and clocks, and (2) how to generate a VHDL test bench

311

from this diagram. We have not demonstrated the timing analysis features of The WaveFormer, such as min/max delay path calculations, setup and hold checking, timing parameter equations, reconvergent fanout calculation, and library generation. These features are covered in the on-line tutorials and examples that ship with the evaluation version.

Appendix E: VHDL Keyword List

Where possible, references to the appropriate IEEE 1076 Language Reference Manual sections (for example, [LRM 7.2]) have been included. Keywords that are new in IEEE 1076-1993 are so indicated.

abs
[Operator] An absolute value operator which can be applied to any numeric type in an expression.

Example:

Delta <= **abs**(A-B)

[LRM 7.2]

access
[Keyword] Declares an access subtype. Access subtypes are used like pointers to refer to other objects. The objects which an access subtype can reference are array objects, record objects, and scalar type objects. An access declaration includes—in this order—the reserved word "access", followed by a subtype.

Example:

```
type AddressPtr is access RAM;
```

[LRM 3.3]

after
[Keyword] Used in signal assignment statements to indicate a delay value before a signal assignment takes place. A signal assignment statement containing an **after** clause includes—in this order—the name of the signal object, the reserved signal assignment symbol "<=", the optional keyword "transport", an expression specifying the value to be assigned to the signal, the reserved word "after", and the delay value (of type "time") after which the signal assignment is to take place.

If no **after** clause is present in a signal assignment statement, an implicit "after 0ns" clause is assumed.

Examples:

```
Clk <= not Clk after 50 ns;
Waveform <= transport '1' after 100 ps;
```

[LRM 8.4]

alias
[Keyword] An alternate name for an object. An alias is primarily used to create a slice (a one-dimensional array referring to all or part) of an existing array. An alias is not a new object, but only an alternate name for all or part of an existing object.

Examples:

```
alias LOWBYTE  :std_logic_vector(7 downto 0) is
        Data1(7 downto 0);
alias HIGHBYTE  :std_logic_vector(7 downto 0) is
        Data1(15 downto 8);
```

Note: aliases cannot be used for multi-dimensional arrays.

[LRM 4.3]

all [Keyword] Used in the following ways:

(1) in a **use** statement, to make all the items in a package visible,

(2) in an attribute specification, to refer to all the names in a name class,

(3) in a configuration specification (for) statement, to refer to all instances of a component, and

(4) in a signal disconnection specification, to refer to all signal drivers of the same type.

Examples:

```
use ieee_std_logic_1164.all;
for DUT: compare use entity work.compare(compare1);
```

[LRM 5.1, 5.2, 5.3]

and [Operator] The logical "and" operator which can be used in an expression. The expression "A and B" returns **true** only if <u>both</u> A and B are true.

[LRM 7.2]

architecture [Keyword] Defines the internal details of a design entity. An architecture body defines the relationships between the input and output elements of the entity. An architecture body consists of a series of concurrent statements. (An architecture body can also include processes, functions, and procedures, each of which may include sequential statements. Although the statements inside a process, for example, are executed sequentially, the process itself is treated within the architecture body as a concurrent statement.)

A given architecture can be associated with only one entity. However, a given entity may have more than one architecture body.

An architecture statement includes—in this order—the following:

(1) the reserved word "architecture", followed by :

    (a) the name of the architecture,

    (b) the reserved word "of",

    (c) the entity name, and

    (d) the reserved word "is",

(2) a declarations section,

(3) the reserved word "begin",

(4) the architecture body (a series of concurrent statements as described above), and

(5) the reserved word "end", followed optionally by the name of the architecture from (1)(a) above.

Example:

```
architecture   sample_architecture of compare is
begin
        GT <= '1' when A > B else '0';
        LT <= '1' when A < B else '0';
        EQ <= '1' when A = B else '0';
end sample_architecture;
```

[LRM 1.2]

array

[Keyword] Defines an array. An array is an object containing a collection of elements that are all of the same type. An array can be either constrained or unconstrained. A constrained array is defined with an index defining the number of array elements. In an unconstrained array, the number of elements in the array is specified in the array's object declaration, or the index definition for the array may be given in a subtype declaration. Arrays may be one-dimensional (single index) or multi-dimensional (multiple indices).

An array definition includes—in this order—the following:

(1) the reserved word "array", followed by a definition(s) of the elements in the array, and

(2) the reserved word "of", followed by the subtype of the array's elements.

Examples:

> **type** DataWord **is array** (15 **downto** 0) **of** DataBit;
> --Constrained

> **type** BigWord **is array** (integer **range** <>) **of** DataBit;
> -- Unconstrained

[LRM 3.2]

assert

[Keyword] A statement which checks to see if a given condition is true. An **assert** statement includes two options, either or both of which may be used:

(1) report—which displays a user-defined message if the given condition is false, and

(2) severity—which allows the user to choose a severity level if the given condition is false.

The four possible severity levels are: **Note, Warning, Error,** and **Failure**. The value of severity is typically used to control the actions of a simulation in the event the given condition is false. For example, a severity level of **Failure** may be used to stop the simulation.

Example:

> **assert** (S = S_expected)
> > **report** "S does not match the expected value!"
> > **severity** Error;

[LRM 8.2, 8.3]

317

attribute   [Keyword] A specification which describes a characteristic of a given object. An attribute is most often used to get additional information about an object. For example, an attribute may be used to find the width of an array or to determine if a signal is in transition (i.e., has an event occurring on it).

Examples:

```
if Clk'event then ...
W = Data'width;
```

An attribute can be a constant, function, range, signal, type, or value. User-defined attributes are always constants, no matter what type. The other five possibilities—function, range, signal, type, and value—are pre-defined attributes.

An attribute declaration is used to declare an attribute name and its type. It includes—in this order—the reserved word "attribute", the name of the attribute, and the attribute's type.

Example:

```
attribute enum_encoding: string;
```

An attribute specification assigns a value to the attribute. It includes—in this order—the reserved word "attribute", the attribute's name, the reserved word "of", an item name (which can be an architecture, component, configuration, constant, entity, function, label, package, procedure, signal, subtype, type, or variable), the name class of the item (e.g., architecture, component, configuration, etc.), the reserved word "is", and an expression.

```
attribute enum_encoding of StateReg is
   "0001 0011 0010 0110 0100 1100 1000";
```

An attribute name must be declared in an attribute declaration before it can be used in an attribute specification.

[LRM 4.4, 5.1, 6.6]

begin

[Keyword] See *block*, *architecture*, *entity*, *if*, and *process*.

block

[Keyword] A concurrent statement used to represent a portion of a design. Block statements may also include an optional Guard feature which allows the user to disable signal drivers within the block when a specified Guard condition is false.

A block statement includes—in this order—the following:

(1) block label,

(2) the reserved word "block",

(3) optionally, a Boolean guard expression (for example, TESTCOUNT<5),

(4) a block header, which specifies the interface of the block with its environment,

(5) a block declarations section,

(6) the reserved word "begin",

(7) the block statements, and

(8) the reserved words "end block", optionally followed by the block label (which, if used, must be the same as the block label in (1)).

When a guard expression is used, a signal driver can be disabled by inserting the reserved word "guarded" at the beginning of the right side of the signal driver statement. For example, based on the example in (3) above, the block statement:

SAMPLE <= **guarded** D;

will cause the signal SAMPLE to take on the value of D only when TESTCOUNT<5. Otherwise, no action on that assignment statement will be taken.

Example:

```
TESTPARITY: block
        signal Atmp,Btmp;  -- Local signals
  begin
        Atmp <= gen_parity(A);
        Btmp <= gen_parity(B);
        ParityEQ <= '1' when Atmp = Btmp else '0';
  end block TESTPARITY;
```

[LRM 9.1]

body

[Keyword] See *package body*.

buffer

[Keyword] One of five possible modes for an interface port. (The other four are **in**, **out**, **inout**, and **linkage**.) The **buffer** mode indicates a port which can be used for both input and output, and it can have only one source. A **buffer** port can only be connected to another **buffer** port or to a signal that also has only one source.

[LRM 1.1]

bus

[Keyword] One of two *kinds* of signals used in a signal declaration (the other is **register**). A **bus** signal represents a hardware bus and defaults to a user-specified value when all of the signal's drivers are turned off.

See also *register*.

[LRM 1.1]

case

[Keyword] A sequential statement used within a process, procedure or function that selects and executes one statement sequence among a list of alternatives, based on the value of a given expression. The expression must be of a discrete type or a one-dimensional array type.

A **case** statement includes—in this order—the following:

(1) the reserved word "case",

(2) the expression to be evaluated,

(3) the reserved word "is",

(4) the reserved word "when" followed by a choice and the sequence of statements to be executed if the expression evaluates to be that choice,

(5) optionally, subsequent "when" statements similar to (4),

(6) optionally, the reserved words "when others" followed by the sequence of statements to be executed if the expression evaluates to be any value other than those specified in the "when" statements above,

(7) the reserved words "end case".

Because the **case** statement chooses one and only one alternative for execution, all possible values for the expression must be covered in "when" statements.

A **case** statement is distinguished from a chain of **if-then-else** statements in that no priority is implied for the conditions specified.

Example:

```
case current_state is
        when IDLE =>
                if start_key = '1' then
                        current_state <= READ0;
                end if;
        when READ0 =>
                current_state <= READ1;
        when READ1 =>
                current_state <= READX;
        when READX =>
                current_state <= WRITE0;
        when WRITE0 =>
                current_state <= WRITEX;
```

```
                               when WRITEX =>
                                   current_state <= IDLE;
               end case;
```

[LRM 8.8]

component   [Keyword] A component declaration used to define the
                       interface to a lower-level design entity. The component may
                       then be included in a component instantiation statement
                       which itself is included in an architecture body, thus allowing
                       one entity to be used as part of another entity. The compo-
                       nent declaration must be placed in the declaration section of
                       the architecture body, or in a package visible to the architec-
                       ture.

                       Example:

```
               component my_adder
                       port(A,B,Cin: in std_ulogic;
                                   Sum,Cout: out std_ulogic);
               end component;
```

                       [LRM 4.5, 9.6]

configuration   [Keyword] A declaration used to create a configuration for an
                       entity. A configuration declaration for a given entity binds
                       one architecture body to the entity and can bind components
                       of architecture bodies within that entity to other entities. In a
                       given configuration declaration for an entity, only one archi-
                       tecture body can be bound to that entity. However, one entity
                       can have many configurations.

                       Example:

```
               configuration this_build of adder is
                       use work.all;
                       for structure
                               for A1,A2,A3: AddBlock
                                       use entity FullAdd(behavior);
                               end for;
```

> **end for**;
> **end** this_build;

[LRM 1.3]

constant

[Keyword] Declares a constant of a type specified in the constant declaration. A constant declaration includes—in this order—the reserved word "constant", the name of the constant, the optional reserved word "in", the type of the constant, and, optionally, an expression for the value of the constant.

If an expression for the value of the constant is not included in the constant declaration, then the constant is referred to as a deferred constant. A deferred constant may only be included in a package declaration, while the complete constant declaration, including the expression which defines its value, must be included in the package body.

Examples:

    constant RESET: std_ulogic_vector(7 downto 0) := "00000000";
    constant PERIOD: time := 80 ns;

[LRM 4.3]

disconnect

[Keyword] Specifies the time delay to disconnect the guarded feature of a signal which is part of a guarded signal statement. See guarded and block. A disconnect statement includes—in this order—the reserved word "disconnect", the name of the guarded signal, the guarded signal's type, the reserved word "after", and a time expression specifying the time after which the guard feature should be disconnected.

In place of the guarded signal's name, the reserved words "others" or "all" may be used. "Others" refers to all other signal statements in the immediately enclosing declarative

323

region which have not been specified in a separate disconnect statement. "All" refers to all other signal statements in the declarative region.

A given signal driver can have only one disconnect statement.

[LRM 5.3]

downto   [Keyword] Used to indicate a descending range in a range statement or other statement which includes a range (for example, an array type declaration). (The reserved word "to" is used to indicate an ascending range.)

See also *range*.

Example:

signal A0,A1: std_logic_vector(15 **downto** 0);

[LRM 6.5]

else   [Keyword] Used to identify the final alternative in an **if** or **when** statement. See *if*, *when*.

elsif   [Keyword] Used to identify an interim alternative in an **if** statement. See *if*.

end   [Keyword] See *architecture, configuration, entity, package, package body, process*.

end block   [Keyword] Used to signify the end of a **block** statement. See *block*.

end case [Keyword] Used to signify the end of a **case** statement. See *case*.

end component [Keyword] Used to signify the end of a **component** declaration. See *component*.

end for [Keyword] Used to signify the end of a **for** statement. See *for*.

end generate [Keyword] Used to signify the end of a **generate** statement. See *generate*.

end if [Keyword] Used to signify the end of an **if** statement. See *if*.

end loop [Keyword] Used to signify the end of a **loop** statement. See *loop*.

end process [Keyword] Used to signify the end of a process. See *process*.

end record [Keyword] Used to signify the end of a **record** statement. See *record*.

end units [Keyword] Used to signify the end of a **units** statement. See *units*.

entity [Keyword] A declaration used to describe the interface of a design entity.

A design entity is an abstract model of a digital system. A design entity includes (1) an entity declaration (which specifies the name of the entity and its interface ports), and (2) at least one architecture body (which models the internal workings of the digital system).

An entity declaration includes—in this order—the reserved word "entity", the entity's name, the reserved word "is", the following optional statements:

(1) the reserved word "generic" followed by a list of generics and their types,

(2) the reserved word "port" followed by a list of interface port names and their types,

(3) any declaration of entity items,

(4) the reserved word "begin" followed by appropriate entity declaration statements, and

(5) non-optionally, the reserved word "end" followed (optionally) by the entity's name.

The ports of an entity are visible within the architecture(s) of the entity, and may be referenced (have their values read, or have values assigned to them, depending on their mode) as signals within the architecture(s).

Declarations made within an entity statement are visible within the corresponding architecture(s).

Example:

```
entity Mux is
        generic(RISE, FALL: time := 0 ns);
        port(A,B: in std_ulogic;
                Sel: in  std_ulogic;
                Y: out std_ulogic);
end Mux;
```

[LRM 1.1]

exit

[Keyword] A sequential statement used in a loop to cause execution to jump out of the loop. An **exit** statement can only be used in a loop and can include an optional **when** condition. An **exit** statement includes—in this order—the reserved word "exit", an optional loop identifier (if no identifier is given, the **exit** statement is applied to the loop in which the **exit** statement occurs), and, optionally, the reserved word "when" followed by the condition under which the exit action is to be executed.

Example:

```
for idx in vectors'range loop
        apply_vector(vec(idx));
        wait for PERIOD;
        if done = '1' then
                exit;
        end if;
end loop;
```

See also *loop*.

[LRM 8.11]

file

[Keyword] Declares a file. A file declaration includes—in this order—the reserved word "file", the name of the file (as used by the program), the subtype indicator (which must define a file subtype), the reserved word "is", on optional mode indicator (which must be either "in" or "out"), and the file's external name (which must be a string expression and is surrounded by quote marks). If no mode is specified, the default is "in".

Example:

```
file vector_file: text is in "VECTOR.DAT";
```

[LRM 4.3]

for

[Keyword] A statement used to identify:

(1) a block specification in a block configuration,

(2) a component specification in a component configuration,

(3) a parameter specification in a generate statement (see generate),

(4) a parameter specification in a loop statement (see loop), or

(5) a time expression in a wait statement (see wait).

See also *block, configuration, loop, generate, wait.*

function

[Keyword] Defines a group of sequential statements that return a single value. A function specification includes—in this order—the reserved word "function", the function's name, a parameter list (which can only include constants and signal objects, and must all be of mode **in**), the reserved word "return", and the type of the value to be returned by the function.

Example:

```
function to_unsigned (a: std_ulogic_vector)
                return integer is
    alias av: std_ulogic_vector (1 to a'length) is a;
    variable ret,d: integer;
begin
    d := 1;
    ret := 0;

    for i in a'length downto 1 loop
        if (av(i) = '1') then
            ret := ret + d;
        end if;
        d := d * 2;
    end loop;
```

>   **return** ret;
>   **end** to_unsigned;

See also *return*.

[LRM 2.1]

generate   [Keyword] Used to do one of the following:

(1) replicate a set of concurrent statements (a for-generation), or

(2) selectively execute a set of concurrent statements if a specified condition is met (an if-generation).

A generate statement used to replicate a set of concurrent statements includes—in this order—the following:

(a) a label for the generate, followed by the reserved word "for", followed by a parameter specification for the "for",

(b) the reserved word "generate",

(c) the concurrent statements to be replicated,

(d) the reserved words "end generate".

A generate statement used to selectively execute a set of concurrent statements includes—in this order—the following:

(a) a label for the generate, followed by the reserved word "if", followed by the condition for the "if",

(b) the reserved word "generate",

(c) the concurrent statements to be selectively executed if the condition in (a) is true,

(d) the reserved words "end generate".

Example:

G: **for** I **in** 0 **to** (WIDTH - 2) **generate**

    -- This generate statement creates the first
    -- XOR gate in the series...

G0: **if** I = 0 **generate**
    X0: xor2 **port map**(A => D(0), B => D(1), Y => p(0));
end generate G0;

    -- This generate statement creates the middle
    -- XOR gates in the series...

G1: **if** I > 0 **and** I < (WIDTH - 2) **generate**
    X0: xor2 **port map**(A => p(i-1), B => D(i+1), Y => p(i));
end generate G1;

    -- This generate statement creates the last
    -- XOR gate in the series...

G2: **if** I = (WIDTH - 2) **generate**
    X0: xor2 **port map**(A => p(i-1), B => D(i+1), Y => ODD);
end generate G2;

end generate G;

[LRM 9.7]

generic

[Keyword] Used in a configuration to define constants whose values may be controlled by the environment. A generic statement includes—in this order—the reserved word "generic", followed by a list of declarations for the generic constants being defined.

Example:

generic(RISE, FALL: time := 0 ns);

See also *entity*.

[LRM 1.1]

generic map

[Keyword] Used to associate values of constants within a block to constants defined outside the block.

For example, suppose a given entity includes an architecture, and the architecture includes a block. A generic map statement could be used to set the value of an entity constant (which was defined by a "generic" statement in the entity declaration), equal to the value of a block constant (which was defined by a "generic" statement in the block). See *generic*.

A generic map statement includes—in this order—the reserved words "generic map" followed by an association list (e.g., "LOCAL => GLOBAL").

Example:

```
U1: And2
        generic map (RISE_TIME => 2 ns, FALL_TIME
                => 2 ns);
        port map (A => IN1, B => IN2, Y => OUT1);
```

[LRM 5.2]

group

[Keyword, 1076-1993] Used to define a group template or specific group. Groups may be used to give a name to a collection of named entities. A group template declaration includes—in this order—the reserved word "group" followed by a group name, the reserved word "is", and a list of classes enclosed in parentheses.

Example:

```
group signal_pair is (signal, signal);     -- a group of two
                                              signals
```

A group declaration includes—in this order—the reserved word "group" followed by a group name, the character ":", a group template name, and a list of named entities enclosed in parentheses.

Example:

>   **group** G1: signal_pair(Clk1,Clk2);

[LRM 4.6]

guarded   [Keyword] Used to limit the execution of a signal statement within a block when the block includes a guard statement. See *block*.

if   [Keyword] A sequential statement used for describing conditional logic.

Example:

```
if A > B then
    Compare <= GT;
elsif A < B then
    Compare <= LT;
else
    Compare <= EQ;
end if;
```

The condition expression of an if statement must be a Boolean logic expression (meaning that it must evaluate to a True or False value).

If statements are sequential, and may only be used in processes, procedures or functions.

See also: elsif, else.

[LRM 8.7]

impure   [Keyword, 1076-1993] Used to declare a function that may return a different value given the same actual parameters, due to side effects. Impure functions have access to a broader class of values than pure fucntions, and can modify objects that are outside their own scope.

Example:

impure function HoldCheck (Clk, Data) **return** Boolean;

[LRM 2.1]

in

[Keyword] A keyword which can be used in two different ways depending on the context:

(1) One of five possible modes for an interface port (the other four are inout, out, buffer, and linkage); the in mode indicates a port which can be used only for input; and

(2) An optional word in a constant declaration. See *constant*.

[LRM 1.1]

inertial

[Keyword, 1076-1993] Used to specify that a delay is inertial. In the absence of an **inertial** or **transport** keyword, the delay is assumed to be inertial.

Example:

Qout <= A **and** B **inertial after** 12 ns;

[LRM 8.4]

inout

[Keyword] One of five possible modes for an interface port. (The other four are **in**, **out**, **buffer**, and **linkage**.) The **inout** mode indicates a port which can be used for both input and output.

[LRM 1.1]

is
[Keyword] See *architecture, case, configuration, entity, file, package, package body, subtype, type.*

label
[Keyword] Used to specify a label name in an attribute statement. See *attribute.*

library
[Keyword] A context clause used to identify libraries from which design units can be referenced. A library clause includes—in this order—the reserved word "library" followed by a list of library logical names. Using a library clause makes a named library visible to the working environment. However, to use a design unit from within that library, a "use" statement must be included specifying the design unit to be used.

All design units automatically include the following implicit library clause:

library STD, WORK;

See also *use.*

[LRM 11.2]

linkage
[Keyword] One of five possible modes for an interface port. (The other four are **in, out, inout,** and **buffer.**) The **linkage** mode indicates a port which can be used for both input and output, and it can only correspond to a signal.

[LRM 1.1]

literal
[Keyword, 1076-1993] Used in group template declarations.

loop

[Keyword] A statement which executes a series of sequential statements multiple times. A **loop** statement can include either (1) a "while" iteration scheme, (2) a "for" iteration scheme, or (3) no iteration scheme.

(1) A loop statement using a "while" iteration scheme includes—in this order—the following:

(a) an optional loop label,

(b) the reserved word "while", followed by the condition which controls whether the series of sequential statements within the loop is executed, followed by the reserved word "loop",

(c) the series of sequential statements to be executed if the condition in (b) evaluates to be True,

(d) the reserved words "end loop", followed by an optional loop label (which, if used, must be the same as the loop label in (a).

Example:

```
while (I < DBUS'length) loop
    ...
    I := I + 1;
end loop;
```

(2) A loop statement using a "for" iteration scheme includes—in this order—the following:

(a) an optional loop label,

(b) the reserved word "for", followed by a parameter specification for the "for", followed by the reserved word "loop",

(c) the series of sequential statements to be executed for the instances defined in the parameter specification in (b),

(d) the reserved words "end loop", followed by an optional loop label, which, if used, must be the same as the loop label in (a).

Example:

```
for I in 0 to DBUS'length - 1 loop
       ...
end loop;
```

(3) A loop statement with no iteration scheme includes—in this order—the following:

(a) an optional loop label,

(b) the reserved word "loop",

(c) the series of sequential statements to be executed,

(d) the reserved words "end loop", followed by an optional loop label, which, if used, must be the same as the loop label in (a).

A **loop** statement with no iteration scheme continues to execute until some action causes execution t ᴄease. This could be done using an "exit" statement, a "next" statement, or a "return" statement within the loop.

Example:

```
loop
            exit when I = DBUS'length;
            I := I + 1;
end loop;
```

[LRM 8.9]

map   [Keyword] See *generic map* and *port map*.

mod

[Operator] The modulus operator which can be applied to integer types. The result of the expression "A mod B" is an integer type and is defined to be the value such that:

(1) the sign of (A mod B) is the same as the sign of B, and

(2) abs (A mod B) < abs (B), and

(3) (A mod B) = (A * (B - N)) for some integer N.

[LRM 7.2]

nand

[Operator] The logical "not and" operator which can be used in an expression. It produces the opposite of the logical "and" operator. The expression "A nand B" returns True only when (1) A is false, or (2) B is false, or (3) both A and B are false.

[LRM 7.2]

new

[Keyword] Used to create an object of a specified type and return an access value that refers to the created object. A new statement includes—in this order—the allocator (which, when evaluated, refers to the created object), followed by the reserved symbol ":=", followed by the reserved word "new", followed by the type of the object being created, and optionally followed by the reserved "new" and an expression for the initial value of the object being created.

[LRM 7.3]

next

[Keyword] A statement allowed within a loop that causes the current iteration of the loop to be terminated and cycles the loop to the beginning of its next iteration.

A **next** statement includes—in this order—the reserved word "next", an optional loop label (which must be the same as the label of the loop in which the **next** statement occurs), and, optionally, the reserved word "when" followed by a condition which, when True, causes the "next" statement to be executed.

If a "when" clause is not included, a "next" statement is executed as soon as it is encountered.

See also *loop*.

[LRM 8.9]

nor

[Operator] The logical "not or" operator which can be used in an expression. It produces the opposite of the logical "or" operator. The expression "A nor B" returns True only when both A and B are false.

[LRM 7.2]

not

[Operator] The logical "not" operator which can be used in an expression. The expression "not A" returns True if A is false and returns False if A is true.

[LRM 7.2]

null

[Keyword] A statement which performs no action. This statement can be used in situations where it is necessary to explicitly specify that no action is needed. For example, a **null** statement may be useful in a **case** statement where all alternatives must be specified but where no action may be required for some alternatives.

Example:

```
D1 <= '0';         -- Default values...
Strobe <= '0';
Rdy <= '0';
```

```
case current_state is
    when S0 =>
        D1 <= '1';
    when S1 =>
        Strobe <= '1';
    when S2 =>
        Rdy <= '1';
    when others =>
        null;
end case;
```

[LRM 8.13]

of

[Keyword] See *architecture, array, configuration, file.*

on

[Keyword] See *wait.*

open

[Keyword] Used in an association list (within a component instantiation statement) to indicate a port that is not connected to any signal.

[LRM 4.3]

or

[Operator] The logical "or" operator which can be used in an expression. The expression "A or B" returns True if (1) A is true, or (2) B is true, or (3) both A and B are true.

[LRM 7.2]

others

[Keyword] Used to specify all remaining elements in:

(1) an element association (in an aggregate),

(2) an attribute specification,

(3) a configuration specification,

(4) a disconnection specification,

(5) case statement, or

(6) a selected assignment

See also select, case.

Examples:

    when others => **null**;

    constant ZERO: std_ulogic_vector (A**'left** to A**'right**) :=
        (**others**=>0);

out   [Keyword] One of five possible modes for an interface port. (The other four are **in, inout, buffer,** and **linkage**.) The **out** mode indicates a port which can be used only for output.

[LRM 1.1]

package   [Keyword] Specifies a set of declarations which can include the following items: aliases, attributes, components, constants, files, functions, types, and subtypes. A package declaration can also include attribute specifications, disconnection specifications, and use clauses. By grouping common declarations in a package declaration, the package declaration can be imported and used in other design units.

Example:

```
package conversions is
    function to_unsigned (a: std_ulogic_vector) return
        integer;
    function to_vector (size: integer; num: integer) return
        std_ulogic_vector;
end conversions;
```

[LRM 2.5]

package body   [Keyword] Specifies the definitions of the various subprograms (components, functions, etc.) that are declared by the package body's associated package declaration. The package body must have the same name as the package declaration. Only one package body can be associated with each package declaration.

Example:

```
package body conversions is
    function to_unsigned (a: std_ulogic_vector) return
        integer is
        ...
    begin
        ...
    end to_unsigned;
    function to_vector (size: integer; num: integer) return
        std_ulogic_vector is
        ...
    begin
        ...
    end to_vector;
end conversions;
```

[LRM 2.6]

port   [Keyword] Used in a configuration to define the input and output ports of an entity. A port statement includes—in this order—the reserved word "port", followed by a list of declarations for the port signals being defined.

Example:

```
entity Mux is
    port(A,B: in std_ulogic;
         Sel: in  std_ulogic;
         Y: out std_ulogic);
end Mux;
```

See also *entity*.

[LRM 1.1]

341

port map [Keyword] Used to associate signals of ports within a block to ports defined outside the block.

For example, suppose a given entity includes an architecture, and the architecture includes a block. A **port map** statement could be used to set the value of an entity port (which was defined by a "port" statement in the entity declaration), equal to the value of a block port (which was defined by a "port" statement in the block). See *port*.

A **port map** statement includes—in this order—the reserved words "port map" followed by an association list (e.g., "LOCAL_PORT => GLOBAL_PORT"). The association list may use positional or named association, as shown in the following examples. Ports may be left unconnected through the use of the **open** keyword.

Examples:

```
U1: And2 port map (IN1, IN2, OUT1);
U1: And2 port map (A => IN1, B => IN2, Y => OUT1);
A18: AddBlk port map (A => A1, B => A1, S => Sum,
        Cout =>open);
```

postponed [Keyword, 1076-1993] Used to declare a process as a postponed process. Postponed processes do not execute until the final simulation cycle at the currently modeled time.

Example:

```
P1: postponed process (D,Snd,Int)
begin
    -- Statements are postponed to end of simulation cycle
end postponed process;
```

procedure [Keyword] Defines a group of sequential statements which are to be executed when the procedure is called. A procedure does not have a return value, but instead can return any number of values (or no values) via its parameter list. Parameters of a procedure must have a mode associated with them (eg. in, out, inout). Values are returned by using mode out or mode inout. A procedure specification includes—in this order—the reserved word "procedure", the procedure name, and a list of the procedure's parameters (which may be constants, signals, or variables, each of whose modes may be in, out, or inout).

Example:

```
procedure dff (signal Clk,Rst,D; in std_ulogic;
            signal Q: out std_ulogic) is
begin
        if Rst <= '1' then
                Q <= '0';
        elsif rising_edge(Clk) then
                Q <= D;
        end if;
end procedure;
```

See also *function*.

[LRM 2.1, 9.3]

process

[Keyword] Defines a sequential process intended to model all or part of a design entity. A **process** statement includes—in this order—an optional sensitivity list, a declarations section, a "begin" statement, the sequential statements describing the operation of the process, and an "end" statement. The sensitivity list identifies signals to which the process is sensitive. Whenever an event occurs on an item in the sensitivity list, the sequential instructions in the process are executed. If no sensitivity list is provided, the process executes until suspended by a **wait** statement.

In addition to signal and variable assignments, the sequential statements in the body of the process can include the following: **assertion, case, exit, if, loop, next, null, procedure, return**, and **wait**.

343

Example:

```
reg: process(Rst,Clk)
        variable Qreg: std_ulogic_vector(0 to 7);
    begin
        if Rst = '1' then   -- Async reset
            Qreg := "00000000";
        elsif rising_edge(Clk) then
            if Load = '1' then
                Qreg := Data;
            else
                Qreg := Qreg(1 to 7) & Qreg(0);
            end if;
        end if;
        Q <= Qreg;
    end process;
```

[LRM 9.2]

pure

[Keyword, 1076-1993] Used to declare a pure function. Pure functions always return the same value for a given set of input actual parameters, and have no side effects. Pure is assumed if there is no **pure** or **impure** keyword.

Example:

pure function HoldCheck (Clk, Data) **return** Boolean;

[LRM 2.1]

range

[Keyword] Used to define a range constraint for a scalar type. A **range** statement includes—in this order—the reserved word "range", the name of the range, and, optionally, two simple expressions for the outer bounds of the range separated by either the reserved word "to" (the ascending direction indicator) or the reserved word "downto" (the descending direction indicator).

Example:

variable Q: integer **range** 0 **to** 15;

[LRM 3.1]

record   [Keyword] Used to declare a record type its corresponding element types. A **record** statement includes—in this order—the following:

(1) the reserved word "record",

(2) an element declaration which includes—in this order—one or more identifiers which share a common subtype, followed by identification of that subtype,

(3) optionally, additional element declarations of the form specified in (2), and

(4) the reserved words "end record".

An element declaration which includes more than one identifier (for example, "COUNT, SUM, TOTAL: INTEGER") is equivalent to a series of single element declarations.

Example:

```
type test_record is record
        CE: std_ulogic;  -- Clock enable
        Set: std_ulogic;
        Din: std_ulogic;
        CRC_Sum: std_ulogic_vector (15 downto 0);
end record;

type test_array is array(positive range <>) of test_record;
```

[LRM 3.2]

register [Keyword] One of two *kinds* of signals used in a signal declaration (the other is **bus**). A **register** signal represents a hardware storage register and defaults to its last driven value when all of the signal's drivers are turned off.

See also *bus*.

[LRM 1.1]

345

reject
[Keyword, 1076-1993] Used to specify the minimum pulse width to propogate in an **after** clause. If no reject time is specified, the specified delay time is assumed for the reject time.

Example:

Q <= Data **reject** 2 ns **after** 7 ns;   -- Delay is 7 ns,
-- reject time is 2 ns

[LRM 8.4]

rem
[Operator] The remainder operator which can be applied to integer types. The result of the expression "A rem B" is an integer type and is defined to be the value such that:

(1) the sign of (A rem B) is the same as the sign of A, and

(2) abs (A rem B) < abs (B), and

(3) (A rem B) = (A - (A / B) * B).

[LRM 7.2]

report
[Keyword] See *assert*.

return
[Keyword] A sequential statement used at the end of a subprogram (a function or procedure) to terminate the subprogram and return control to the calling object.

When used in a procedure, the reserved word "return" appears alone.

When used in a function, the reserved word "return" must be followed by an expression which defines the result to be returned by the function. The expression's type must be the same type as specified by the return statement in the function's specification. See *function*. A **return** statement must be the last statement executed in a function.

See also *function*, *procedure*.

[LRM 2.1]

rol

[Operator, 1076-1993] Rotate left operator.

Example:

Sreg <= Sreg **rol** 2;

[LRM 7.2.3]

ror

[Operator, 1076-1993] Rotate right operator.

Example:

Sreg <= Sreg **ror** 2;

[LRM 7.2.3]

select

[Keyword] A concurrent signal assignment statement which selects and assigns a value to a target signal from among a list of alternatives, based on the value of a given expression.

A select statement includes—in this order—the following:

(1) the reserved word "with", followed by the expression to be evaluated, followed by the reserved word "select",

(2) the target signal, followed by the reserved symbol "<=", followed by:

(a) the first value which could be assigned to the target signal, followed by the reserved word "when", followed by a choice which, if the expression evaluates to be that choice, will cause the first value to be assigned to the target signal, and

347

(b) second and subsequent values which could be assigned to the target signal, each followed by the reserved word "when", and each followed by a choice which, if the expression evaluates to be that choice, will cause the value to be assigned to the target signal.

Since the select statement chooses one and only one alternative for execution at a given time, all possible values for the expression must be covered in "when" statements. An "others" clause may be used to cover values not explicitly named.

Example:

```
architecture concurrent of mux is
begin
        with Sel select
              Y <= A when "00",
                   B when "01",
                   C when "10",
                   'X' when others;
        end concurrent;
```

[LRM 9.5]

severity   [Keyword]: See *assert*.

signal   [Keyword] Declares a signal of a specified type. A signal declaration includes—in this order—the reserved word "signal", the name of the signal, the subtype of the signal, an optional indication of the signal's kind (which must be either "register" or "bus"), and optionally, an expression specifying the initial value of the signal.

Signals declared within an entity are visible in the corresponding architecture(s).

Note:
A signal cannot be declared within a process, procedure or function.

Example:

```
architecture behavior of fsm is
        signal current_state: state;
        signal DataBuf: std_logic_vector(15 downto 0);
    begin
        ...
    end behavior;
```

See also *guarded* and *block*.

[LRM 4.3]

sla   [Operator, 1076-1993] Shift left arithmetic operator.

Example:

Addr <= Addr **sla** 8;

[LRM 7.2.3]

sll   [Operator, 1076-1993] Shift left logical operator.

Example:

Addr <= Addr **sll** 8;

[LRM 7.2.3]

sra   [Operator, 1076-1993] Shift right arithmetic operator.

Example:

Addr <= Addr **sra** 8;

[LRM 7.2.3]

srl   [Operator, 1076-1993] Shift right logical operator.

Example:

$$Addr <= Addr \ \textbf{srl} \ 8;$$

[LRM 7.2.3]

subtype   [Keyword] Declares a subtype (a type with a constraint that is based on an existing parent type). A subtype declaration includes—in this order—the reserved word "subtype", the subtype's identifier, the reserved word "is", an optional resolution function, the base type of the subtype, and an optional constraint. If no constraint is included, the subtype is the same as the specified base type.

Examples:

```
subtype short is integer range 0 to 255;
subtype X01Z is std_ulogic range 'X' to 'Z';
```

[LRM 4.2]

then   [Keyword] Part of the syntax of an **if** statement. See **if**.

to   [Keyword] Used to indicate an ascending range in a **range** statement or other statement which includes a range (for example, a **variable** statement). (The reserved word "downto" is used to indicate a descending range.)

See also *range*.

Example:

```
signal A0,A1: std_ulogic_vector(0 to 15);
```

[LRM 6.5]

transport   [Keyword] Used to specify non-inertial delay in a signal assignment statement.

Examples:

Waveform <= **transport** '1' **after** 10 ns;

See also *after*.

[LRM 8.4]

type   [Keyword] Declares a type. There are two kinds of type declarations: a full type declaration and an incomplete type declaration.

A full type declaration includes—in this order—the reserved word "type", the type identifier, the reserved word "is", and the type definition. A type definition can be an access type, a composite type, a file type, or a scalar type.

An incomplete type declaration includes only the reserved word "type" followed by the type's identifier. If an incomplete type declaration exists, a full type declaration with the same identifier must also exist. The full type declaration must occur after the incomplete type declaration and within the same declarations section as the incomplete type declaration.

Note that two type declarations define two different types, even if the definitions are the same and they differ only by their respective identifiers.

Examples:

```
type StateMachine is (RESET, IDLE, READ, WRITE,
      ERROR);
type RAD16 is range 0 to 15;
type test_record is record
     CE: std_ulogic;  -- Clock enable
     Set: std_ulogic;
     Din: std_ulogic;
     CRC_Sum: std_ulogic_vector (15 downto 0);
```

end record;

[LRM 4.1]

unaffected

[Keyword, 1076-1993] Used to indicate in a conditional or selected signal assignment when the signal is not to be given a new value.

Example:

```
Mux <= A when Sel = "00" else
       B when Sel = "01" else
       C when Sel = "10" else
       unaffected;
```

[LRM 8.4]

units

[Keyword] Used in a type declaration to declare physical types. A **units** statement includes—in this order:

(1) the reserved word "units",

(2) the base unit,

(3) optionally, one or more secondary units, and

(4) the reserved words "end units".

Example:

```
type time  is range -2_147_483_647 to 2_147_483_647
    units
        fs;
        ps  = 1000 fs;
        ns  = 1000 ps;
        us  = 1000 ns;
        ms  = 1000 us;
        sec = 1000 ms;
        min = 60 sec;
        hr  = 60 min;
    end units;
```

[LRM 3.1]

until   [Keyword]: See *wait*.

use   [Keyword] A context clause used to identify items in other design units so those items can be referenced. A **use** clause includes—in this order—the reserved word "use", followed by a list of design units (or design unit items) to be referenced.

A **use** clause makes the referenced design units visible to the working environment. If a design unit (or design unit item) belongs to a library different from the current library, a **library** statement must be included before the **use** statement. The **library** statement must specify the library holding the referenced design unit.

> Examples:
>
> > **use** mylib.mypackage.dff;
> > **use** mylib.mypackage.**all**;
> > **use** mylib.**all**;
> > **use** work.**all**;

All design units automatically include the following two implicit clauses:

> > **library** STD, WORK;
> > **use** STD.STANDARD.**all**;

See also *library*.

[LRM 10.4]

variable   [Keyword] Declares a variable of a specified type. A variable declaration includes—in this order—the reserved word "variable", the variable's name, the variable's subtype, and, optionally, an expression specifying the initial value of the variable.

A variable can only be declared within a process, procedure or function. Also, a variable cannot be of a file type. Variables declared within a process have their values preserved during subsequent executions of the process. Variables declared within a function or procedure have their values initialized each time the function or procedure is called.

Example:

```
process(Rst,Clk)
    variable Q: integer range 0 to 15;
begin
    if Rst = '1' then              -- Asynchronous reset
        Q := 0;
    elsif rising_edge(Clk) then
        if Load = '1' then
            Q := to_unsigned(Data);   -- Convert vector to
                                      -- integer
        elsif Q = 15 then
            Q := 0;
        else
            Q := Q + 1;
        end if;
    end if;
    Count <= to_vector(4,Q);       -- Convert integer to
                                   -- vector
end process;
```

[LRM 4.3]

wait

[Keyword] A statement used to temporarily suspend a process until:

(1) a specified time has passed ("wait for", followed by a time expression), or

(2) a specified condition is met ("wait until", followed by a Boolean expression), or

(3) an event occurs which affects one or more signals ("wait on", followed by a sensitivity list which specifies signals on each of which an event must occur before processing continues).

When a **wait** statement is used within a process, the process must not include a sensitivity list.

Example:

```
CLOCK: process
    variable c: std_ulogic := '0';
    constant PERIOD: time := 50 ns;
begin
    wait for PERIOD / 2;
    c := not c;
    clk <= c;
end process;
```

[LRM 8.1]

when

[Keyword] Used to specify a condition during which an **exit** or **next** statement will be executed. See *exit* and *next*. Also used to specify a choice (or choices) within a **case** statement. See *case*.

while

[Keyword] Used to specify a condition during which a loop will be executed. See *loop*.

with

[Keyword] Part of the syntax of a selected signal assignment. See *select*.

xnor

[Operator, 1076-1993] The logical "both or neither" (equality) operator which can be used in an expression. The expression "A xnor B" returns True only when (1) A is true and B is true, or (2) A is false and B is false.

[LRM 7.2]

xor [Operator] The logical "one or the other but not both" (in-equality) operator which can be used in an expression. The expression "A xor B" returns True only when (1) A is true and B is false, or (2) A is false and B is true.

[LRM 7.2]

= [Operator] The equality operator which can be used in an expression on any type except file types. The resulting type of an expression using this operator is Boolean (that is, True or False). The expression "A = B" returns True only if A and B are equal.

[LRM 7.2]

/= [Operator] The inequality operator which can be used in an expression on any type except file types. The resulting type of an expression using this operator is Boolean (that is, True or False). The expression "A /= B" returns True only if A and B are not equal.

[LRM 7.2]

:= [Operator] The assignment operator for a variable. The expression "TEST_VAR := 1" means that the variable TEST_VAR is assigned the value 1.

[LRM 8.5]

< [Operator] The "less than" operator which can be used in an expression on scalar types and discrete array types. The resulting type of an expression using this operator is Boolean (that is, True or False). The expression "A < B" returns True only if A is less than B.

[LRM 7.2]

<=

[Operator] This symbol has two purposes. When used in an expression on scalar types and discrete array types, it is the "less than or equal to" operator. The resulting type of an expression using this operator in this context is Boolean (that is, True or False). In this context, the expression "A <= B" returns True only if A is less than or equal to B.

Example:

LE := '1' **when** A <= B **else** '0';

In a signal assignment statement, the symbol "<=" is the assignment operator. Thus, the expression "TEST_SIGNAL <= 5" means that the signal TEST_SIGNAL is assigned the value 5.

Example:

DataBUS <= 0x"E800";

[LRM 7.2, 8.4]

>

[Operator] The "greater than" operator which can be used in an expression on scalar types and discrete array types. The resulting type of an expression using this operator is Boolean (that is, True or False). The expression "A > B" returns True only if A is greater than B.

[LRM 7.2]

>=

[Operator] The "greater than or equal to" operator which can be used in an expression on scalar types and discrete array types. The resulting type of an expression using this operator is Boolean (that is, True or False). The expression "A >= B" returns True only if A is greater than or equal to B.

[LRM 7.2]

+

[Operator] The addition operator. Both operands must be numeric and of the same type. The result is also of the same numeric type. Thus, if A = 2 and B = 3, the result of the expression "A + B" is 5.

This operator may also be used as a unary operator representing the identity function. Thus, the expression "+A" would be equal to A.

[LRM 7.2]

-

[Operator] The subtraction operator. Both operands must be numeric and of the same type. The result is also of the same numeric type. Thus, if A = 5 and B = 3, the result of the expression "A - B" is 2.

This operator may also be used as a unary operator representing the negative function. Thus, the expression "-A" would be equal to the negative of A.

[LRM 7.2]

&

[Operator] The concatenation operator. Each operand must be either an element type or a 1-dimensional array type. The result is a 1-dimensional array type.

[LRM 7.2]

*

[Operator] The multiplication operator. Both operands must be of the same integer or floating point type.

The multiplication operator can also be used where one operand is of a physical type and the other is of an integer or real type. In these cases, the result is of a physical type.

[LRM 7.2]

/ [Operator] The division operator. Both operands must be of the same integer or floating point type.

The division operator can also be used where a physical type is being divided by either an integer type or a real type. In these cases, the result is of a physical type. Also, a physical type can be divided by another physical type, in which case the result is an integer.

[LRM 7.2]

** [Operator] The exponentiation operator. The left operand must be of an integer type or a floating point type, and the right operand (the exponent) must be of an integer type. The result is of the same type as the left operand.

[LRM 7.2]

359

Appendix F: Driving Game Listings

Driving Game (Chapter 9)

The following VHDL source files describe the driving game that was discussed in Chapter 9, *Writing Test Benches*. The driving game was originally inspired by a design example written in Altera's AHDL and published in the Altera Max+Plus II Getting Started manual. This example is written using VHDL 1076-1993 language features.

PIZZAPAK.VHD

```
-----------------------------------------------------------------------
-- Package game_types contains declarations for the Direction and Speed
-- data types and corresponding constant values. This package is to be
-- compiled into library 'work'.
--
library ieee;
use  ieee.std_logic_1164.all;

package game_types is
    subtype tDirection is std_logic_vector(1 downto 0);
    subtype tSpeed is std_logic;
    constant FAST: tSpeed := '1';
    constant SLOW: tSpeed := '0';
    constant NORTH: tDirection := "00";
    constant EAST: tDirection := "01";
```

```
        constant WEST: tDirection := "10";
        constant SOUTH: tDirection := "11";
end game_types;
```

PIZZATOP.VHD

```
-------------------------------------------------------------------------
-- Pizzatop is a top-level module that connects the two state machines
-- (navigate and trap) together with two instances of a 4-bit counter to
-- create the complete game design. Refer to the block diagram of
-- chapter 9 for a graphic view of the component interconnections.
--
library ieee;
use  ieee.std_logic_1164.all;

use work.game_types.all;

use work.counter;
use work.navigate;
use work.trap;

entity pizzatop is
    port(Clk: in std_logic;
        Reset: in std_logic;
        Speed: in tSpeed;
        Dir: in tDirection;
        Party: out std_logic;
        Tickets: out std_logic_vector(3 downto 0);
        DriveTime: out std_logic_vector(3 downto 0));
end entity pizzatop;

architecture structure of pizzatop is
    component counter is
        port(Increment: in std_logic;
            Reset: in std_logic;
            Clk: in std_logic;
            Count: out std_logic_vector(3 downto 0));
    end component counter;
    component navigate is
        port(Dir: in tDirection;
            Speed: in tSpeed;
            Reset: in std_logic;
            Clk: in std_logic;
            Party: out std_logic;
```

```vhdl
            GetTicket: out std_logic;
            Speeding: out std_logic);
    end component navigate;
    component trap is
        port(Speeding: in std_logic;
             Reset: in std_logic;
             Clk: in std_logic;
             GetTicket: out std_logic);
    end component trap;

    signal TicketDetect: std_logic;
    signal Ticket1: std_logic;
    signal Ticket2: std_logic;
    signal Speeding: std_logic;
    signal ResetBar: std_logic;
begin
    -- Instantiate the components...
    COUNT1: counter port map(Increment=>TicketDetect, Reset=>Reset,
Clk=>Clk, Count=>Tickets);
    COUNT2: counter port map(Increment=>ResetBar, Reset=>Reset,
Clk=>Clk, Count=>DriveTime);
    NAV1: navigate port map(Dir=>Dir, Speed=>Speed, Reset=>Reset,
Clk=>Clk, Party=>Party, GetTicket=>Ticket2, Speeding=>Speeding);
    TRAP1: trap port map(Speeding=>Speeding, Reset=>Reset, Clk=>Clk,
GetTicket=>Ticket1);

    ResetBar <= not Reset;
    TicketDetect <= Ticket2 or Ticket1;

end architecture structure;
```

PIZZACTL.VHD

```vhdl
--------------------------------------------------------
-- This file contains the "guts" of the design: the two state
-- machine controllers (navigate and trap) and the counter block.
--
library ieee;
use ieee.std_logic_1164.all;
use work.game_types.all;


--------------------------------------------------------
```

```
-- Navigation state machine.
-- This state machine defines the town map, which is represented
-- here as a state machine. The map defines 7 locations, and we
-- have also added a "Lost" state to handle the situation where our
-- hero has taken a wrong turn. For each state (location on the map)
-- we define the resulting next state for a given speed and direction
-- combination.
--
entity navigate is
    port (Dir: in tDirection;
        Speed: in tSpeed;
        Reset,Clk: in std_logic;
        Party: out std_logic;
        GetTicket, Speeding: out std_logic);
end entity navigate;

architecture behavior of navigate is
    type state_machine is (Office, Ramp1, Ramp2, Ramp3, Downtown,
                    Beach, Pizza, Lost);
    signal present_state, next_state: state_machine;
    signal have_pizza: std_logic;
begin

    -- State registers...
    registers: process(reset,clk)
    begin
        if reset = '1' then
            present_state <= Office;
            have_pizza <= '0';
        elsif rising_edge(clk) then
            present_state <= next_state;
            if present_state = Pizza then
                have_pizza <= '1';
            end if;
        end if;
    end process registers;

    -- Transitions...
    -- The descriptions of the transitions are simplified somewhat through
    -- the use of a procedure (drive) that accepts the next state for each
    -- of the two possible speeds for a given direction. This procedure
    -- does double-duty by checking for speeding on residential and
    -- commercial streets and setting the speeding and warning flags
    -- appropriately.
    transitions: process(present_state,dir,speed,have_pizza)
        type road is (Residential, Commercial, Expressway, OffRoad);
```

```
procedure drive(dest1,dest2: state_machine;
        road_type: road) is
begin
  if speed = SLOW then
    next_state <= dest1;
    speeding <= '0';
    GetTicket <= '0';
  else
    next_state <= dest2;
    if road_type = Residential then
      speeding <= '1';
      GetTicket <= '1';
    elsif road_type = Commercial then
      speeding <= '1';
      GetTicket <= '0';
    else            -- Expressway (or lost)
      speeding <= '0';
      GetTicket <= '0';
    end if;
  end if;
end procedure drive;

begin
  case present_state is
  when Office =>
    party <= '0';
    if Dir = NORTH then
      drive(Ramp1,Ramp2,Expressway);
    else
      drive(Lost,Lost,OffRoad);
    end if;
  when Ramp1 =>
    party <= '0';
    if Dir = NORTH then
      drive(Ramp2,Ramp3,Expressway);
    elsif Dir = EAST then
      drive(Downtown,Beach,Residential);
    else
      drive(Lost,Lost,OffRoad);
    end if;
  when Ramp2 =>
    party <= '0';
    if Dir = WEST then
      drive(Ramp1,Ramp1,Expressway);
    elsif Dir = EAST then
      drive(Ramp3,Ramp3,Expressway);
```

```
        elsif Dir = SOUTH then
            drive(Downtown,Downtown,Residential);
        else
            drive(Lost,Lost,OffRoad);
        end if;
    when Ramp3 =>
        party <= '0';
        if Dir = WEST then
            drive(Ramp2,Ramp1,Expressway);
        elsif Dir = SOUTH then
            drive(Beach,Pizza,Commercial);
        else
            drive(Lost,Lost,OffRoad);
        end if;
    when Downtown =>
        party <= '0';
        if Dir = NORTH then
            drive(Ramp2,Ramp2,Residential);
        elsif Dir = EAST then
            drive(Beach,Beach,Residential);
        elsif Dir = WEST then
            drive(Ramp1,Ramp1,Residential);
        else
            drive(Lost,Lost,OffRoad);
        end if;
    when Beach =>
        party <= have_pizza;
        if Dir = NORTH then
            drive(Ramp3,Ramp3,Commercial);
        elsif Dir = WEST then
            drive(Downtown,Ramp1,Residential);
        elsif Dir = SOUTH then
            drive(Pizza,Pizza,Commercial);
        else
            drive(Lost,Lost,OffRoad);
        end if;
    when Pizza =>
        party <= '0';
        if Dir = NORTH then
            drive(Beach,Ramp3,Commercial);
        else
            drive(Lost,Lost,OffRoad);
        end if;
    when Lost =>
        party <= '0';
        drive(Lost,Lost,OffRoad);
```

```vhdl
        end case;
      end process transitions;
   end architecture behavior;


--------------------------------------------------------
-- Speed trap state machine.
-- This state machine keeps track of how many warnings the
-- driver has had. The machine starts out in state "legal", meaning
-- that no warnings or tickets have been accumulated. If the driver
-- is found to be speeding at any time (as determined in the navigate
-- state machine), this machine advances to state 'warning'. Any
-- further infraction will result in the machine advancing to state
-- 'ticket'.
library ieee;
use  ieee.std_logic_1164.all;
use  work.game_types.all;

entity trap is
   port (Speeding,Reset,Clk: in std_logic;
         GetTicket: out std_logic);
end entity trap;

architecture behavior of trap is
   type state_machine is (legal, warn, ticket);
   signal present_state, next_state: state_machine;
   signal gt: std_logic;
begin

   registers: process(reset,clk)
   begin
     if reset = '1' then
        present_state <= legal;
        GetTicket <= '0';
     elsif rising_edge(clk) then
        present_state <= next_state;
        GetTicket <= gt;
     end if;
   end process registers;

   transitions: process(present_state,speeding)
   begin
     case present_state is
     when legal =>
        if speeding = '1' then
           next_state <= warn;
           gt <= '0';
```

```vhdl
        else
           next_state <= legal;
           gt <= '0';
        end if;
      when warn =>
        if speeding = '1' then
           next_state <= ticket;
           gt <= '1';
        else
           next_state <= warn;
           gt <= '0';
        end if;
      when ticket =>
        next_state <= legal;
        gt <= '0';
      end case;
    end process transitions;
end architecture behavior;


-----------------------------------------------------------------
-- 4-bit counter. This counter is described using the numeric_std
-- (1076.3) data type 'unsigned'.
--
library ieee;
use ieee.std_logic_1164.all;
use ieee.numeric_std.all;

entity counter is
   port(Increment,reset,clk: in std_logic;
      count: out std_logic_vector(3 downto 0));
end entity counter;

architecture behavior of counter is
   signal q: unsigned(3 downto 0);
begin
   cnt1: process(reset,clk)
   begin
     if reset = '1' then
        q <= (others => '0');
     elsif rising_edge(clk) then
        if increment = '1' then
           if q = 15 then
              q <= (others => '0');
           else
              q <= q + 1;
           end if;
```

```
            end if;
          end if;
       end process cnt1;
       count <= STD_LOGIC_VECTOR(q);
    end architecture behavior;
```

Appendix G: Synopsys Textio Package

The following package declaration describes the Synopsys **std_logic_textio** package. The complete source code for this package, including its package body, can be found on the companion CD-ROM in the file **\EXAMPLES\VHDL93\ADD-SUB\SYNTEXT.VHD**.

```
----------------------------------------------------------------------------
--
-- Copyright (c) 1990, 1991, 1992 by Synopsys, Inc.  All rights reserved.
--
-- This source file may be used and distributed without restriction
-- provided that this copyright statement is not removed from the file
-- and that any derivative work contains this copyright notice.
--
--      Package name: STD_LOGIC_TEXTIO
--
--      Purpose: This package overloads the standard TEXTIO procedures
--               READ and WRITE.
--
--      Author: CRC, TS
--
----------------------------------------------------------------------------
```

```
use STD.textio.all;
library IEEE;
use  IEEE.std_logic_1164.all;

package STD_LOGIC_TEXTIO is
--synopsys synthesis_off
        -- Read and Write procedures for STD_ULOGIC and
STD_ULOGIC_VECTOR
        procedure READ(L:inout LINE; VALUE:out STD_ULOGIC);
        procedure READ(L:inout LINE; VALUE:out STD_ULOGIC;
                GOOD: out BOOLEAN);
        procedure READ(L:inout LINE; VALUE:out
                STD_ULOGIC_VECTOR);
        procedure READ(L:inout LINE; VALUE:out
                STD_ULOGIC_VECTOR; GOOD: out BOOLEAN);
        procedure WRITE(L:inout LINE; VALUE:in STD_ULOGIC;
                JUSTIFIED:in SIDE := RIGHT; FIELD:in WIDTH := 0);
        procedure WRITE(L:inout LINE; VALUE:in
                STD_ULOGIC_VECTOR;
                JUSTIFIED:in SIDE := RIGHT; FIELD:in WIDTH := 0);

        -- Read and Write procedures for STD_LOGIC_VECTOR
        procedure READ(L:inout LINE; VALUE:out
                STD_LOGIC_VECTOR);
        procedure READ(L:inout LINE; VALUE:out
                STD_LOGIC_VECTOR; GOOD: out BOOLEAN);
        procedure WRITE(L:inout LINE; VALUE:in
                STD_LOGIC_VECTOR;
                JUSTIFIED:in SIDE := RIGHT; FIELD:in WIDTH := 0);

        --
        -- Read and Write procedures for Hex and Octal values.
        -- The values appear in the file as a series of characters
        -- between 0-F (Hex), or 0-7 (Octal) respectively.
        --

        -- Hex
        procedure HREAD(L:inout LINE; VALUE:out
                STD_ULOGIC_VECTOR);
        procedure HREAD(L:inout LINE; VALUE:out
                STD_ULOGIC_VECTOR; GOOD: out BOOLEAN);
        procedure HWRITE(L:inout LINE; VALUE:in
                STD_ULOGIC_VECTOR;
                JUSTIFIED:in SIDE := RIGHT; FIELD:in WIDTH := 0);
        procedure HREAD(L:inout LINE; VALUE:out
```

```
                STD_LOGIC_VECTOR);
procedure HREAD(L:inout LINE; VALUE:out
                STD_LOGIC_VECTOR; GOOD: out BOOLEAN);
procedure HWRITE(L:inout LINE; VALUE:in
                STD_LOGIC_VECTOR;
                JUSTIFIED:in SIDE := RIGHT; FIELD:in WIDTH := 0);

-- Octal
procedure OREAD(L:inout LINE; VALUE:out
                STD_ULOGIC_VECTOR);
procedure OREAD(L:inout LINE; VALUE:out
                STD_ULOGIC_VECTOR; GOOD: out BOOLEAN);
procedure OWRITE(L:inout LINE; VALUE:in
                STD_ULOGIC_VECTOR;
                JUSTIFIED:in SIDE := RIGHT; FIELD:in WIDTH := 0);
procedure OREAD(L:inout LINE; VALUE:out
                STD_LOGIC_VECTOR);
procedure OREAD(L:inout LINE; VALUE:out
                STD_LOGIC_VECTOR; GOOD: out BOOLEAN);
procedure OWRITE(L:inout LINE; VALUE:in
                STD_LOGIC_VECTOR;
                JUSTIFIED:in SIDE := RIGHT; FIELD:in WIDTH := 0);


--synopsys synthesis_on
end STD_LOGIC_TEXTIO;
```

Appendix H: Glossary

Access Type

A data type analogous to a pointer that provides a form of data indirection. An access value is returned by an *allocator*.

Actual Parameter

An object or literal being passed as an argument to a subprogram, or being used as a higher-level port or generic in a component instantiation.

Aggregate

A form of expression that denotes the value of a *composite type* (such as an array or record). An aggregate value is specified by listing the value of each element of the aggregate, using either *positional* or *named association*.

Alias

Used to declare an alternate name for an object. Often used in subprograms to "normalize" the direction or range of an array argument.

Allocator

An operation in VHDL that creates an anonymous variable object. Allocators return *access values* that may be used to access the variable object.

Architecture

A *design unit* that describes the actual function (operation) of all or part of your design. An architecture must be associated with (bound to) an *entity*. All VHDL design descriptions must include at least one architecture.

Architecture Body

That portion of an *architecture* declaration existing between the **begin** and **end** statements of the architecture.

ASIC

Short for Application-Specific Integrated Circuit. A circuit designed for a specific application, rather than being an off-the-shelf component.

Assert

A statement that checks whether a given condition is true. If the condition is not true, a report (message) is generated during simulation.

Attribute

A special identifier used to return or specify information about a named entity. Predefined attributes are prefixed by the ' character. Other, non-standard attributes may be defined for specific VHDL design tools.

Array

A collection of one or more elements of the same type. Array data types are declared with an array range and an array element type, and may have more than one dimension.

Base Type

The type on which a subtype is based. For example, an array subtype may be defined with a constrained range, and be based on another array type that is unconstrained.

Binding

The association of a specific component instance with a lower-level entity and architecture. Binding may be specified using configuration statements or specifications, or may be implied by the default binding.

Block

A VHDL feature allowing partitioning of the design description within an architecture.

Compile

The process of analyzing VHDL source file statements to create an intermediate form. In the PeakVHDL simulation environments, the compilation process results in 32-bit Windows object files that must be linked to create a simulation executable.

Component

An entity that has been declared for use in a higher-level design entity.

Component Declaration

A statement defining the interface (port list and optional generic list) to an entity that is to be instantiated in the current design entity.

Component Instantiation

A concurrent statement that references a declared component and creates one unique instance of that component. A component instantiation includes an instance name, a component (entity) name, an optimal generic map, and a port map.

Composite Type

A data type, such as an *array* or *record*, that includes more than one constituent *element*.

Concurrent

A characteristic of statements within the *architecture body* of a design description. Concurrent statements have no order dependency, and describe operations that are inherently parallel.

Configuration Statement

An optional design unit that specifies how a project is to be assembled prior to simulation. Configurations are somewhat akin to a parts list for your design and specify such things as the binding of entities to architectures, the mapping of components and ports to their lower-level equivalents, etc.

Constant

An declared object that has a constant value and cannot be changed. Constants are used to give symbolic names to literal values, and may be declared globally (within packages) or locally.

Constraint

A finite range of possible values for a type or subtype.

Declared Entity

An object, type, subprogram or other element of the design that has been declared and given an identifying name. Declared entities have scoping, meaning that they are not visible outside the scope in which they were declared.

Declaration

A statement entered in a declarative region of the design description (such as in a package, or prior to the **begin** statement of an entity, architecture, block, process or subprogram) that creates a *declared entity*.

Delta Cycle

A simulation cycle in which all non-postponed processes and other concurrent statements are repetitively executed and signal assignments are scheduled until no more events are pending. A complete delta cycle occurs in zero simulated time (the start and end time of the delta cycle are the same).

Descending Range

A range that is specified with the **downto** keyword.

Design Entity

The combination of an entity and its corresponding architecture. The minimum VHDL design description includes at least one design entity.

Design Unit

A separately compilable section of VHDL source code. The five types of design units are entities, architectures, packages, package bodies and configurations. Each design unit must have a unique name within the project.

Driver

The combination of a given *signal* and its current and projected future values. When two or more drivers exist for a given signal (such as when multiple values are specified for the signal at the same point in time), a *resolution function* is required.

Element

One entry in a composite type, such as an array. A one-dimensional array declared with the array bounds **0 to 7**, for example, would have eight elements.

Entity

A *design unit* that describes the interface (inputs and outputs) of all or part of your design. All VHDL design descriptions must have at least one entity declaration.

Enumeration Literal

A symbolic representation of an enumerated type value. Enumeration literals may take the form of either identifiers or characters.

Enumerated Type

A symbolic data type that is declared with an enumerated type name, and one or more enumeration values (*enumeration literals*).

Event

A change in the value of a signal at a given point in simulated time. Events are scheduled (rather than immediate), and always occur at the beginning of the next *delta cycle*. Events can also be delayed through the use of the **after** keyword.

Exit Condition

A expression combined an **exit** statement that specifies a condition under which a *loop* should be terminated.

Expression

A syntactically-correct sequence of operators, keywords and literals that defines some computed value.

Field Name

An identifier that provides access to one *element* of a *record* data type.

File Type

A data type that represents an arbitrary-length sequence of values of a given type. The most typical application of a file type is to represent a disk file, such as might be read or written during simulation.

For Loop

A *loop* construct in which the iteration scheme is a **for** statement. The **for** statement specifies a precise, finite range of the loop, and creates an index variable for the loop.

Formal Parameter

An identifier used within a subprogram declaration or other context in which actual parameters, ports or generics are to be later substituted.

Generic

A parameter passed to an entity, component or block that describes additional, instance-specific information about that entity, component or block.

Global Declaration

A declaration that is visible to multiple design entities, as in the case of a declaration made within a package.

Hierarchy

The structure of a design description, expressed in terms of a tree of related components. Most simulated designs include at least two levels of hierarchy: the test bench and the lower-level design description.

Identifier

A sequence of characters that uniquely identify a *named entity* in a design description.

Index

A scalar value that specifies a precise element, or range of elements, within an array.

Infinite Loop

A loop that has no iteration scheme. Infinite loops must include one or more exit conditions to avoid

Iteration Scheme

The start and exit conditions for a *loop*. The iteration scheme is expressed using a **for** or **while** statement in a loop. A loop with no iteration scheme is an *infinite loop.*

Library

A storage facility allowing one or more VHDL source files (and their corresponding design units) to be placed in a common location. Libraries are referenced in a VHDL source file through the use of the **library** statement.

Literal

A specific value that can be applied to an object of a some type. Literals fall into five general catagories: bit strings, enumeration literals, numeric literals, strings, or the special literal **null**.

Loop

A sequential state providing the ability to repeat one or more statements. A loop may be finite (as in the case of a **for** loop) or infinite, depending on the nature of its *iteration scheme.*

LRM

Language Reference Manual. The official document, numbered 1076, that defines the precise syntax and semantics of VHDL.

Mode

The direction of data (either **in, out, inout** or **buffer**) of a *subprogram parameter* or *entity port*.

Named Association

A method of explicitly associating actual *parameters* and other designators with formal designators, as in a *subprogram* reference or *component instantiation*.

Named Entity

A item that has been declared and given a unique name (within the current *scope*). Examples of named entities include such things as *signals* and *variables, entities, architectures* and *blocks, component instances, processes, loop* labels, etc.

Object

A named entity that can be assigned (or initialized with) a value and that has a specific type. Objects include signals, constants, variables and files.

Parameter

An *object* or *literal* passed into a subprogram via that subprogram's parameter list.

Physical Types

A data type used to represent measurements. A physical type value is specified with an integer literal followed by a unit that has been defined for the type. One example of a standard physical type is the type **time**, which has the units **fs, ps, ns, us, ms, sec, min,** and **hr.**

Port

A interface element of an *entity* or *component*. A port must be specified with a name, data type and *mode*.

Positional Association

A method of specifying the mapping of *actual* and *formal parameters* (or actual and formal *ports*) by position in a list.

Programmable Logic

A device technology allowing the function of a digital circuit to be programmed into a blank chip using an array of programmable fuses, static RAM, or other means.

Process

A collection of *sequential* statements that are executed whenever there is an *event* on any *signal* appearing in the process *sensitivity list* or, in the case in which there is no sensitivity list, whenever an event occurs (or simulation time is reached) that satifies the condition of a *wait* statement within the process. Signals assigned within a process are not updated until the current *delta cycle* is complete.

Range

A subset of the possible values of a *scalar* type. Ranges may be used, for example, to specify a *type* or *subtype* declaration, or to specify the range of a *loop*.

Record

A data type that includes more than one *element* of differing types. Record elements are identified by *field names*.

Resolution Function

A special function, associated with a *type declaration*, that describes the resulting value when two or more different values are driven onto a signal of that data type.

Scalar

A data type that has a distinct order to its values, allowing two objects or literals of that type to be compared using relational operators. Scalar types include *integers, enumerated types, floating point types*, and *physical types*.

Sensitivity List

A list of signals associated with a process, specifying under what conditions (in terms of events) the process is to be executed. In the absence of a sensitivity list, the process must include one or more **wait** statements.

Sequential

A characteristic of statements within *processes* and *subprograms*. Sequential statements are executed in sequence, and therefore have order dependency. Sequential statements may be used to describe sequential logic (logic that implies memory elements), or may be used to describe combinational (non-sequential) logic.

Signal

An storage object that maintains a history of *events*. Signals are created as the result of *signal declarations* or *port declarations*.

Signal Declaration

A statement that introduces (creates) a new signal. Signals are declared with a name and a data type. Signals may be declared globally (in a *package*) or locally (such as in the declarative region of an *entity*, *architecture*, or *block*). Assignments to signals are scheduled, meaning that do not have new values assigned to them until the current *delta cycle* has completed.

Slice

A one-dimensional, contiguous *array* created as a result of constraining a larger one-dimensional array.

Source File

A file containing VHDL statements describing all or part of your design.

String

A string data type is an *array* of characters. String data types may be constrained (fixed length) or unconstrained. A string *literal* is a sequence of characters enclosed by two quotation marks (").

Subprogram

A function or procedure. Subprograms may be declared globally (in a package) or locally (such as in the declarative region of an entity, architecture, block, process or subprogram).

Test Bench

One or more VHDL source files describing the sequence of tests to be run on your design. Test benches are normally the top-level of a design being simulated.

Transaction

A scheduled (current or future) *event* for a given *signal*. A transaction consists of a value and a time at which that value is to be driven onto the signal.

Time Unit

A symbolic unit of time used during simulation. Time units are defined as part of the VHDL specification and include fs (femtoseconds), ps (picoseconds), ns (nanoseconds), us (microseconds), ms (milliseconds), sec (seconds), min (minutes), and hr (hours).

Type

A declared name and a corresponding set of declared values representing the possible values of the type. Types fall into the following general catagories: *scalar types, composite types, file types*, and *access types*.

Type Conversion

An operation that results in the conversion of one data type to another. Type conversions may be implicit, explicit, or make use of type conversion functions.

Type Declaration

A *declaration statement* that creates a new data type. A type declaration must include a type name and a description of the entire set of possible values for that type.

Variable

A storage facility used in *processes* and *subprograms* to represent local data. Variables must be declared with a name and *type*. Variables are persistent with processes (meaning they retain their values in subsequent executions of the process), but are non-persistent in subprograms.

Waveform

A series of *transactions* defining the behavior of a *signal* (in terms of future *events*) over time.

Appendix I: Other Resources

Standards Documents

The following standards documents are available from the IEEE, and are "must haves" for serious VHDL users:

Standard 1076-1993, *IEEE Standard VHDL Language Reference Manual*, IEEE, 1994.

Standard 1164-1993, *IEEE Standard Multivalue Logic System for VHDL Model Interoperability*, IEEE, 1993.

Standard 1076.3, *VHDL Synthesis Packages*, IEEE, 1995.

Standard 1076.4, *VITAL ASIC Modeling Specification*, IEEE, 1995.

To order IEEE documents, contact the IEEE at the following address:

IEEE Service Center
445 Hoes Lane
P.O. Box 1331
Piscataway, NJ 08855-1331 USA

Phone: 1-800-678-IEEE

In Europe, use the following address:

IEEE Computer Society
13 Avenue de l'Aquilon
B-1200 Brussels BELGIUM

Phone: 32.2.770.21.98

Other VHDL Texts

There are many books in addition to this one on the subject of VHDL. The following list is just a sampling:

Armstrong, James R. and F. Gail Gray, *Structured Logic Design With VHDL*, Prentice Hall, Englewood Cliffs, N.J., 1993.

Ashenden, Peter, *The Designer's Guide to VHDL*, Morgan Kaufman Publishers, 1993.

Bhasker, Jayaram, *A VHDL Primer*, Prentice Hall, Englewood Cliffs, NJ, 1992.

Bhasker, Jayaram, *A VHDL Synthesis Primer*, Star Galaxy Publishing, Allentown, PA, 1996.

Coelho, David, *The VHDL Handbook*, Kluwer Academic Publishers, Hingham, MA, 1989.

Lipsett, Roger, Carl Schaeffer and Cary Ussery, *VHDL: Hardware Description and Design*, Kluwer Academic Publishers, Hingham, MA, 1989.

Mazor, Stanley and Patricia Langstraat, *A Guide to VHDL*, Kluwer Academic Publishers, Hingham, MA, 1991.

Perry, Douglas L., *VHDL*, 2nd edition, MacGraw-Hill, Inc., New York, NY, 1993.

Rushton, Andrew, *VHDL for Logic Synthesis*, McGraw-Hill, New York, NY, 1995.

VHDL International

VHDL International is an organization dedicated to the promotion and ongoing evolution of VHDL. The organization holds conferences twice yearly, and publishes a newletter for VHDL users and developers. Contact VHDL International at the following address:

VHDL International
407 Chester Street
Menlo Park, CA 94025
(800) 554-2550
(415) 329-0758

On-line Resources

Newsgroups and Web sites related to VHDL have existed for some time, and new sites are constantly appearing. The following pointers will help to get you started:

Newsgroups

comp.lang.vhdl

comp.lsi.cad

sci.electronics.cad

Web Sites

The following commercial Web sites include VHDL-related information and resources, or they belong to companies that offer VHDL-related design tools and services.

http://www.acc-eda.com/

The Accolade Design Automation, Inc. Home Page includes an on-line VHDL tutorial, as well as information about Accolade's PeakVHDL and PeakFPGA products.

http://www.actel.com/

Actel, Inc. produces FPGA devices and offers VHDL-related design software for their devices.

http://www.altera.com/

Altera Corporation also produces FPGA devices, and their MAX+PLUS software includes a VHDL input option.

http://www.amd.com/

Advanced Micro Devices produces complex PLDs and offers third-party VHDL synthesis support for their devices.

http://www.atmel.com/

ATMEL, Inc. is another producer of complex PLDs and also resells third-party VHDL synthesis tools.

http://www.attme.com/

AT&T Microelectronics (now Lucent Technologies) offers FPGA devices, as well as design tools.

http://www.cadence.com/

Cadence Design Systems is a major vendor of VHDL and Verilog-related products.

http://www.capilano.com/

Capilano Computing produces the DesignWorks schematic entry and simulation product, and offers VHDL-related products as well.

http://www.chrysalis.com/

Chrysalis is a producer of VHDL and Verilogic-related formal verification tools.

http://www.cypress.com/

Cypress Semiconductor produces complex PLD and FPGA devices, and offers a low-cost ($99) VHDL synthesis package.

http://www.data-io.com/

Data I/O Corporation and its Synario division offer design tools for FPGAs and complex PLDs.

http://www.escalade.com/

Escalade is a producer of VHDL-related design tools.

http://www.exemplar.com/

Exemplar Logic is a producer of VHDL and Verilog synthesis products for FPGA and gate array devices.

http://www.ikos.com/

IKOS, Inc. produces hardware accelerated simulation products.

http://www.latticesemi.com/

Lattice Semiconductor Corporation is a producer of FPGA and complex PLD devices, and offers design tools in support of those devices.

http://www.libtech.com/

Library Technologies is a vendor of simulation and synthesis libraries.

http://www.mentor.com/

Mentor Graphics is a major vendor of electronic design tools, including VHDL simulation and synthesis tools.

http://www.model.com/

Model Technology produces the V-System simulator, which supports both VHDL and Verilog.

http://www.orcad.com/

OrCAD is a vendor of personal computer-based design tools and offers VHDL simulation and synthesis tools.

http://www.quicklogic.com/

Quicklogic produces FPGA devices and offers VHDL synthesis and simulation tools in support of its devices.

http://www.synopsys.com/

Synopsys is a major vendor of simulation and synthesis tools based on both VHDL and Verilog.

http://www.synplicity.com/

Synplicity offers FPGA synthesis tools that accept both VHDL and Verilog.

http://www.syncad.com/

SynaptiCAD produces The Waveformer, a timing diagram analysis tool that is capable of generating VHDL and Verilog test benches from waveforms.

http://www.veribest.com/

Veribest (formerly Intergraph Electronics) produces a wide range of electronic design automation tools.

http://www.viewlogic.com/

Viewlogic produces electronic design automation tools, including VHDL and Verilog simulation and synthesis products, that operate on personal computers and workstations.

http://www.xilinx.com/

Xilinx products FPGA and complex PLD devices, and offers VHDL design tools in support of those devices.

Index

C

function 97
function kind array attribute 102–103
function kind attribute 101–102, 103–105, 105–106
prefix 97
range kind attribute 108–119
signal 97
signal kind attribute 106–108
type 97
type kind attribute 108
value 97

L

Label
  instance 151
  keyword label 334
  loop labels 198
Language
  C 16
  human language specification 33
  Language Reference Manual 313
  netlist 3
  PLD 17
  programming 3, 36
  programming language interface 16
  proprietary 10–13
  standard 4
  structural 3
  test 3
  VHSIC Hardware Description Language 4
Large projects 8
'Last_active 103–105, 105
'Last_event 103–105, 104, 105
'Last_value 97, 103–105, 104, 105
Latch 172
  unintended latch 172
  unintended latches 274–275
  unwanted 146
Latches
  implied 45
Layout 12
Learning
  curve 13–14
'Left 98, 98–100, 102, 102–103
'Leftof 101–102, 102

Length
  pulse 158
'Length 98, 98–100
Library 32, 384
  ASIC libraries 278
  default 215
  design libraries 217–219
  ieee 43, 129
  keyword library 334
  named libraries 218
  standard 31, 89, 111
  statement 43, 232
  Text I/O library 89
  unit 25
  user-defined 32
  VITAL-compliant libraries 280
  work library 218, 232
Limitations
  of synthesis 265
Linkage
  keyword linkage 334
Linked list 89
List
  argument 199
  generic 26, 224
  keyword 313–359
  linked 89
  parameter 56
  parts 29, 226
  port 26, 224
  processes with sensitivity lists 166–168
  processes without sensitivity lists 168–179
  sensitivity 48, 51–59, 166
  sensitivity list 165, 387
Literal 67, 74–77, 384
  ambiguous literal type 114–119
  based 76–77
  bit string 74–75
  character 74, 82
  enumeration literal 381
  integer 75
  keyword literal 334
  numeric 75–76
  physical 77–78
  real 75
  string 74

LICENSE AGREEMENT AND LIMITED WARRANTY

READ THE FOLLOWING TERMS AND CONDITIONS CAREFULLY BEFORE OPENING THIS CD PACKAGE. THIS LEGAL DOCUMENT IS AN AGREEMENT BETWEEN YOU AND PRENTICE-HALL, INC. (THE "COMPANY"). BY OPENING THIS SEALED CD PACKAGE, YOU ARE AGREEING TO BE BOUND BY THESE TERMS AND CONDITIONS. IF YOU DO NOT AGREE WITH THESE TERMS AND CONDITIONS, DO NOT OPEN THE CD PACKAGE. PROMPTLY RETURN THE UNOPENED CD PACKAGE AND ALL ACCOMPANYING ITEMS TO THE PLACE YOU OBTAINED THEM FOR A FULL REFUND OF ANY SUMS YOU HAVE PAID.

1.　**GRANT OF LICENSE:** In consideration of your purchase of this book, and your agreement to abide by the terms and conditions of this Agreement, the Company grants to you a nonexclusive right to use and display the copy of the enclosed software program (hereinafter the "SOFTWARE") on a single computer (i.e., with a single CPU) at a single location so long as you comply with the terms of this Agreement. The Company reserves all rights not expressly granted to you under this Agreement.

2.　**OWNERSHIP OF SOFTWARE:** You own only the magnetic or physical media (the enclosed CD) on which the SOFTWARE is recorded or fixed, but the Company and the software developers retain all the rights, title, and ownership to the SOFTWARE recorded on the original CD copy(ies) and all subsequent copies of the SOFTWARE, regardless of the form or media on which the original or other copies may exist. This license is not a sale of the original SOFTWARE or any copy to you.

3.　**COPY RESTRICTIONS:** This SOFTWARE and the accompanying printed materials and user manual (the "Documentation") are the subject of copyright. The individual programs on the CD are copyrighted by the authors of each program. Some of the programs on the CD include separate licensing agreements. If you intend to use one of these programs, you must read and follow its accompanying license agreement. You may *not* copy the Documentation or the SOFTWARE, except that you may make a single copy of the SOFTWARE for backup or archival purposes only. You may be held legally responsible for any copying or copyright infringement which is caused or encouraged by your failure to abide by the terms of this restriction.

4.　**USE RESTRICTIONS:** You may *not* network the SOFTWARE or otherwise use it on more than one computer or computer terminal at the same time. You may physically transfer the SOFTWARE from one computer to another provided that the SOFTWARE is used on only one computer at a time. You may *not* distribute copies of the SOFTWARE or Documentation to others. You may *not* reverse engineer, disassemble, decompile, modify, adapt, translate, or create derivative works based on the SOFTWARE or the Documentation without the prior written consent of the Company.

5.　**TRANSFER RESTRICTIONS:** The enclosed SOFTWARE is licensed only to you and may *not* be transferred to anyone else without the prior written consent of the Company. Any unauthorized transfer of the SOFTWARE shall result in the immediate termination of this Agreement.

6.　**TERMINATION:** This license is effective until terminated. This license will terminate automatically without notice from the Company and become null and void if you fail to comply with any provisions or limitations of this license. Upon termination, you shall destroy the Documentation and all copies of the SOFTWARE. All provisions of this Agreement as to warranties, limitation of liability, remedies or damages, and our ownership rights shall survive termination.

7.　**MISCELLANEOUS:** This Agreement shall be construed in accordance with the laws of the United States of America and the State of New York and shall benefit the Company, its affiliates, and assignees.

8.　**LIMITED WARRANTY AND DISCLAIMER OF WARRANTY:** The Company warrants that the SOFTWARE, when properly used in accordance with the Documentation, will operate in substantial conformity with the description of the SOFTWARE set forth in the Documentation. The Company does not warrant that the SOFTWARE will meet your requirements or that the operation of the SOFTWARE will be uninterrupted or error-free. The Company warrants that the media on which the SOFTWARE is delivered shall be free from defects in materials and workmanship under normal use for a period of thirty (30) days from the date of your purchase. Your only remedy and the Company's only obligation under these limited warranties is, at the Company's option, return of the warranted item for a refund of any amounts paid by you or replacement of the item. Any replacement of SOFTWARE or media under the warranties shall not extend the original warranty period. The limited warranty set forth above shall not apply to any SOFTWARE which the Company determines in good faith has been subject to misuse, neglect, improper installation, repair, alteration, or damage by you. EXCEPT FOR THE EXPRESSED WARRANTIES SET FORTH ABOVE, THE COMPANY DISCLAIMS ALL WARRANTIES, EXPRESS OR IMPLIED, INCLUDING WITHOUT LIMITATION, THE IMPLIED WARRANTIES OF MERCHANTABILITY AND FITNESS FOR A PARTICULAR PURPOSE. EXCEPT FOR THE EXPRESS WARRANTY SET FORTH ABOVE, THE COMPANY DOES NOT WARRANT, GUARANTEE, OR MAKE ANY REPRESENTATION REGARDING THE USE OR THE RESULTS OF THE USE OF THE SOFTWARE IN TERMS OF ITS CORRECTNESS, ACCURACY, RELIABILITY, CURRENTNESS, OR OTHERWISE.

IN NO EVENT, SHALL THE COMPANY OR ITS EMPLOYEES, AGENTS, SUPPLIERS, OR CONTRACTORS BE LIABLE FOR ANY INCIDENTAL, INDIRECT, SPECIAL, OR CONSEQUENTIAL DAMAGES ARISING OUT OF OR IN CONNECTION WITH THE LICENSE GRANTED UNDER THIS AGREEMENT, OR FOR LOSS OF USE, LOSS OF DATA, LOSS OF INCOME OR PROFIT, OR OTHER LOSSES, SUSTAINED AS A RESULT OF INJURY TO ANY PERSON, OR LOSS OF OR DAMAGE TO PROPERTY, OR CLAIMS OF THIRD PARTIES, EVEN IF THE COMPANY OR AN AUTHORIZED REPRESENTATIVE OF THE COMPANY HAS BEEN ADVISED OF THE POSSIBILITY OF SUCH DAMAGES. IN NO EVENT SHALL LIABILITY OF THE COMPANY FOR DAMAGES WITH RESPECT TO THE SOFTWARE EXCEED THE AMOUNTS ACTUALLY PAID BY YOU, IF ANY, FOR THE SOFTWARE.

SOME JURISDICTIONS DO NOT ALLOW THE LIMITATION OF IMPLIED WARRANTIES OR LIABILITY FOR INCIDENTAL, INDIRECT, SPECIAL, OR CONSEQUENTIAL DAMAGES, SO THE ABOVE LIMITATIONS MAY NOT ALWAYS APPLY. THE WARRANTIES IN THIS AGREEMENT GIVE YOU SPECIFIC LEGAL RIGHTS AND YOU MAY ALSO HAVE OTHER RIGHTS WHICH VARY IN ACCORDANCE WITH LOCAL LAW.

ACKNOWLEDGMENT

YOU ACKNOWLEDGE THAT YOU HAVE READ THIS AGREEMENT, UNDERSTAND IT, AND AGREE TO BE BOUND BY ITS TERMS AND CONDITIONS. YOU ALSO AGREE THAT THIS AGREEMENT IS THE COMPLETE AND EXCLUSIVE STATEMENT OF THE AGREEMENT BETWEEN YOU AND THE COMPANY AND SUPERSEDES ALL PROPOSALS OR PRIOR AGREEMENTS, ORAL, OR WRITTEN, AND ANY OTHER COMMUNICATIONS BETWEEN YOU AND THE COMPANY OR ANY REPRESENTATIVE OF THE COMPANY RELATING TO THE SUBJECT MATTER OF THIS AGREEMENT.

Should you have any questions concerning this Agreement or if you wish to contact the Company for any reason, please contact in writing Robin Short, Prentice Hall PTR, One Lake Street, Upper Saddle River, New Jersey 07458.